Integrated Life Cycle Design of Structures

T0187631

Current objectives towards sustainable building and civil engineering are leading to new challenges for the construction industry. The life cycle engineering approach takes into account all aspects of construction practice, from design, construction, and service life management, to demolition and the recycling of materials.

Traditionally the process of design has concentrated on the construction phase itself, with the primary object being to optimise efficiency and minimise costs during development and construction. With the move towards more sustainable development comes the need for this short-term approach to be expanded to encompass the entire service life of the structure.

This book explains how to optimise structures for their entire design life, through an optimum integrated life cycle design process. Sustainability and performance issues are detailed.

Integrated Life Cycle Design of Structures provides a comprehensive account of this rapidly emerging field. It is essential reading for civil and structural engineers, designers, architects, contractors, and clients.

Asko Sarja is Research Professor of Building Technology in Structural Engineering, Technical Research Centre of Finland (VTT).

Integrated Life Cycle Design of Structures

Asko Sarja

CRC Press
Taylor & Francis Group
Boca Raton London New York

CRC Press is an imprint of the
Taylor & Francis Group, an **informa** business
A SPON PRESS BOOK

CRC Press
Taylor & Francis Group
6000 Broken Sound Parkway NW, Suite 300
Boca Raton, FL 33487-2742

First issued in paperback 2019

ISBN-13: 978-0-415-25235-5 (hbk)
ISBN-13: 978-0-367-45477-7 (pbk)

Typeset in Times by
HWA Text and Data Management, Tunbridge Wells

British Library Cataloguing in Publication Data
A catalogue record for this book is available from the British Library

Library of Congress Cataloging in Publication Data
Sarja, Asko.
    Integrated life cycle design of structures / Asko Sarja.
        p. cm.
    Includes bibliographical references and index.
        1. Buildings, 2. Structural engineering. 3. Product life cycle.
    4. Building materials–Service life. 5. Sustainable development.
    I. Title.

TH845 .S35 2002
624.1–dc21

                                                                    2001049380

Visit the Taylor & Francis Web site at
http://www.taylorandfrancis.com

and the CRC Press Web site at
http://www.crcpress.com

Contents

Tables

Figures

Photograph

Preface

Our society is living through a period of great change, in which we can also see changes in the central goals and requirements of construction techniques. The challenge to the present generation is to lead rapid development of a global economy towards sustainability in relation to our entire society, economy, social welfare and ecology.

Buildings, and civil and industrial infrastructures are the longest lasting and most important products of our society. The economic value contained in buildings, and civil and industrial infrastructures are, to say the least, significant; and the safe, reliable and sound economic and ecological operation of these structures is greatly needed. In industrialised countries buildings and civil infrastructures represent about 80 per cent of national property. Construction plays a major role in the use of natural resources and in the development of the quality of the natural environment in our time. Consequently, building and civil engineering can make a major contribution to the sustainable development of society.

The sustainability of buildings and built environment can, in short, be defined as thinking in time spans of several generations. Sustainability includes social aspects (welfare, health, safety, comfort), economic aspects, functional aspects (usability for changing needs), technical aspects (serviceability, durability, reliability) and ecological aspects (consumption of natural resources such as energy, raw materials and water; air, water and soil pollution, waste production; and impact on biodiversity), all treated over the entire life cycle of the built facilities. It could be claimed that a built facility can only be as good as its design. I have proposed a technical definition for sustainable building as: 'Sustainable building is a technology and practice which meets the multiple requirements of the people and society in an optimal way during the life cycle of the built facility' (Sarja, 1997).

Design is an important part of construction: translating the requirements of owners, users and society into performance requirements of the structural system; creating and optimising structural solutions which fulfil those requirements and finally, proving through analysis and dimensioning calculations, that these requirements are fulfilled.

The new model of integrated life cycle design presented in this book includes a framework for integrated structural life cycle design, a description of the design process and its phases, special life cycle design methods regarding different aspects like life cycle costs, life cycle environmental impact, design for reuse and recycling, service life design, durability design, multiple attribute optimisation and multiple attribute decision-making.

This book provides methods and methodologies for design of structures in order to meet the requirements of sustainable developments during the entire service life of the structures. The scope of this book covers both the loadbearing and non-loadbearing structures of buildings, bridges, towers, dams and other structural facilities. This book includes:

- definitions and terms;
- framework schedule of the integral structural design;
- systematics and central methods of the environmental structural design; and
- examples of applications .

Target groups and users of the book are expected to include: architects, structural engineers, civil engineers, clients of construction, contractors, suppliers of materials, components and modules and standards organisations. The methodology and the methods present here are well suited for the use in the development of new building concepts and new structural solutions and products.

The origin for this kind of integrated life cycle design was the idea of the author who concluded that the sustainability aspects have to be quantified and calculation methods have to be developed (Sarja 1994, Sarja 1995). This concept was then realised in the work of the international working committee: RILEM TC 172 EDM / CIB TG22 'Environmental Design Methods in materials and structural engineering', which was active in years 1996–2000 (Sarja 2000). The committee consisted of twenty-two members and corresponding members under the chairmanship of the author, with Mr. Julian Kümmel as the Secretary of the group. I expect the results of the commission's work to be an active starting point towards the final formulation of life cycle design. Chapter 4 by Christoph Müller, Germany and Chapter 5 by Toshio Fukushima, Japan, originated from the commission's report. After preparing the report of RILEM/CIB working group, the results were published in an article in the RILEM journal *Materials and Structures* in 1999 (Sarja 1999). I have used the working group report as a core of this book, but have developed it further and added a considerable amount of new material from results of my other work and from other sources, and have totally reorganised the content into its current form.

Integrated life cycle design is still at a phase of rapid development, and this is the start of the final formulation of a new integrated design process and methodology, which in future will serve as a general design culture. The next step on this path will be an integration of the design, management and maintenance planning of buildings and civil infrastructures into a comprehensive life time engineering. The practical implementation can be carried out through international and national standards, guidelines and computer tools.

Asko Sarja
Espoo, Finland, June 2001

References

Sarja, A. (1994) 'Development towards the ecological and recyclable building materials technology'. Second International Conference on Materials Engineering for Resources, Akita, Japan, 19–22 October , 1994. Research Institute of Natural Resources attached to Mining College, Akitsa University, Japan.

Sarja, A. (1995) 'Methods and methodology on environmental aspects of building materials and structures'. RILEM Workshop on environmental aspects of building materials and structures, 21–22 September. Technical Research Centre of Finland: Espoo.

Sarja, A. (1997) 'A vision of sustainable materials and structural engineering'. In: Tuutti K. (ed.) *Selected Research Studies from Scandinavia.* Report TVBM-3078. Lund University, Lund Institute of Technology: Lund.

Sarja, A. (1999) 'Environmental design methods in materials and structural engineering – Progress Report of RILEM TC 172 EDM / CIB TG22'. *Materials and Structures*, 32, (December) 699–707.

Sarja, A. (ed.) (2000) 'Environmental design methods in materials and structural engineering: Integrated Life Cycle Design', RILEM TC EDM / CIB TG22. Technical Research Centre of Finland: Espoo.

Introduction

Traditionally, design has concentrated on the construction phase of optimising construction costs and short-term performance. Sustainable development gives rise to a need for integrated life cycle design, where all solutions are optimised for the entire design service life of the building. In the building, only the loadbearing frame, and most of the envelope are designed to resist degradation over the design service life. The parts having a shorter design service life, typically the building services, partitions and finishings, will probably be renewed several times during the design service life of the building. For these parts with shorter life cycles, recycling is both economically and ecologically important (Sarja 1997a). The other challenge for design is to guarantee the performance and durability of technical systems over the design service life. For this aim, durability design methods are important. Resistance design is expanded into durability design to include time as a new dimension in the design calculations (Sarja and Vesikari 1996). Health and safety aspects are generally related to the control of moisture and temperature conditions and to special areas such as hazardous emissions from materials.

The definition already includes the principle that sustainability must always be treated according to the life cycle principle. In other words, the application of life cycle methodology to design, manufacture, construction, maintenance, and the management of building projects by companies and other organisations involved in building (Sarja 1997b). With regard to the design of materials and structures, this leads to the idea of integrated life cycle design, which is the subject of this book.

The core competence and working area of structural engineers in the context of sustainable construction is the design of structures for sustainability over their entire life cycles. This means that the structural design process must be looked at anew. Furthermore, new methodologies and calculation methods must be adopted, for example from mathematics, physics, systems engineering and other natural and engineering sciences. However, we have to keep in mind the need of strong methodologies, and transparency and simplicity of the design process and its methods in order to keep these multiple issues under control and to avoid excessive design work.

In life cycle design, analysis and design are expanded into two economic areas: financial costs and environmental costs. Life cycle costs are calculated as either present value or as annual costs by discounting the costs from manufacture, construction, maintenance, repair, changes, modernisation, rehabilitation, re-use, recycling and disposal. Monetary costs are treated, as usual, by current value calculations. Environmental costs are the use of non-renewable natural resources (materials and energy), and the production of air, water or soil pollution. The consequences of air pollution are health problems, inconvenience for people, ozone depletion and global warming. These impacts dictate the environmental profiles of the structural and building service systems. The goal is to limit the environmental costs to permitted values and to minimise them. Integrated lifetime design is an important link in construction: translating the requirements of owners, users and society into performance requirements of the technical systems; creating and optimising technical solutions, which fulfil those requirements; and proving through analysis and dimensioning calculations that the performance requirements will be fulfilled over the entire design service life. The adoption of these new methods and processes will increase the need for renewed education and training of all those involved. This new model of integrated *life cycle design*, also called *lifetime design*, includes a framework for integrated structural life cycle design, a description of the design process and its phases, and special lifetime design methods with regard to different aspects discussed above.

Quality assurance has been widely systematised under the ISO 9000 standards. An environmental efficiency procedure is presented in the ISO 14000 standards. The impact of life cycle principles in construction is in the application of life cycle criteria in the quality assurance procedure. Multiple life cycle criteria are also applied during the selection of products, although most of the product specifications have already been produced at the design phase.

Integrated life cycle design supports an improved quality approach, which can be called life cycle quality (see Figure 0.1). All the areas shown in Figure 0.1 are treated over the life cycle of structures, and controlled in the design by technical performance parameters.

The life cycle performance of structures is highly dependent on maintenance. The first important instructions for life cycle maintenance are produced during the design stage. The structural system of a building or civil engineering facility needs a users' manual, just like a car or any other piece of equipment. The manual will be produced gradually during the design process in co-operation with those involved in design, manufacture and construction. The usual tasks of the structural designer are: compiling a list of maintenance tasks for the structural system, compiling and applying operational instructions, control and maintenance procedures and works, checking and co-ordinating the operational, control and maintenance instructions of product suppliers and contractors, preparing the relevant parts of the users' manual, and checking relevant parts of the final users' manual.

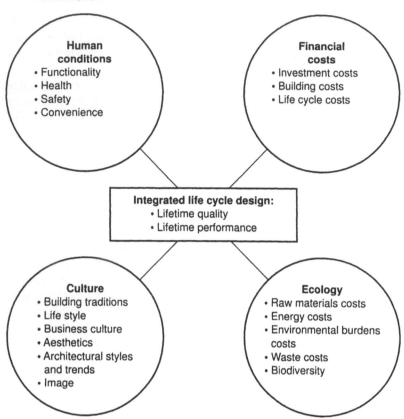

Figure 0.1 Main aspects of life cycle quality (Sarja 1996, Sarja 1999)

   The active reduction of waste during construction, renovation and demolition is possible through the selective dismantling of structural systems, components and materials specifically for recycling. Selective dismantling includes detailed planning of the dismantling phases, and optimising the work sequences and logistics of the dismantling and selection process. The main goal is to separate the different types of materials and different types of components at the demolition phase in order to avoid multiple actions. The recyclability of the building materials and structural components depends on the degree and/or technical level of the desired reuse.

References

Sarja, A. (1996) 'Environmental design methods in materials and structural engineering'. *CIB Information*, 4(96), 23–25.

Sarja, A. (1997a) 'A vision of sustainable materials and structural engineering', in Tuutti K. (ed.) *Selected Research Studies from Scandinavia*. Report TVBM-3078. Lund University, Lund Institute of Technology: Lund.

Sarja, A. (1997b) 'Framework and methods of life cycle design of buildings', Symposium: Recovery, Recycling, Reintegration, R'97, 4–7 February, Geneva. *EMPA*, VI, 100–105.

Sarja, A. (1999) 'Environmental design methods in materials and structural engineering – Progress Report of RILEM TC 172-EDM / CIB TG22'. *Materials and Structures*, 32 (December), 699–707.

Sarja, A. and Vesikari, E. (1996) *Durability Design of Concrete Structures*. RILEM Report Series 14. E&FN Spon: London.

1 Framework of integrated structural life cycle design

1.1 Scope and objectives

Integrated structural design provides methods and methodologies for structural design to meet the requirements of sustainable development during the entire service life of structures, and thus to achieve a good life cycle quality. The scope of this book includes both loadbearing and non-loadbearing structures of buildings, bridges, towers, dams, harbours, tunnels and other structural facilities. This book includes:

- definitions and terms
- framework schedule of the integral structural design
- systematics and central methods of the environmental structural design
- examples of applications in building and bridge design.

Architects, structural engineers, civil engineers, construction clients, and standards organisations are the target readers for this book.

The objective for the development of environmentally-oriented design is to realise design methods and methodologies for structural design in order to meet the requirements of sustainable development during the entire life cycle of the structures (resources, transport, manufacture, use, recycling and reuse, demolition, waste disposal). Environmental design will be presented as a part of integrated structural design, which includes the mechanical, physical, economic, energy, health and safety, and environment aspects. Integrated design will manage these multiple requirements in a systematic way.

1.2 Standards and codes related to integrated life cycle design

A number of general principles, and assessment and analysis methods, which serve as a basis for development of integrated life cycle design already exist. However, there is no consistent methodology which is only focused on the design of materials and structures, which is the subject of this book.

The general framework of environmental management is defined in ISO 14000 standards: ISO 14001: 'Environmental Management Systems. Specification with guidance for use' and ISO 14004: 'Environmental Management Systems. General guidelines on principles, systems and supporting techniques'.

The role of this book is to provide design tools for fulfilment of the structural design issues defined in ISO 14004, Chapter 4.1 'Commitment and policy' and Chapter 4.2 'Planning'. Product information about building materials and building components serve as an important information source of structural design. The following draft standards and committee drafts can be applied when evaluating environmental product information: ISO DIS 14040: 'Environmental Management – Life Cycle Assessment – Principles and Framework' and ISO 14041: 'Environmental Management – Life Cycle Assessment – Inventory analysis'.

Technical performance principles, and service life planning and prediction are treated in Draft International Standard ISO/DIS 15686-1.

National methodology codes exist in several countries. Often their scope is more general than just building technology, but the methodologies are applicable with some modifications also in the building sector. Some national codes, standards and guidelines are listed in Appendixes 2 and 3. Comprehensive guidelines for integrated life cycle design of structures have been published by the Association of Finnish Civil Engineers (Sarja 2000b, 2001).

1.3 General features of environmentally efficient structural technology and design

In ISO 14001 and 14004 the principle of a continued improvement process to achieve improvements in overall environmental performance is expressed. The environment is defined as the surroundings in which an organisation operates (including air, water, land, natural resources, flora, fauna, humans), and their interaction. Ecological aspects are the elements of an organisation's activities, products or services that interact with the environment. Environmental impact is any change to the environment, whether adverse or beneficial, wholly or partially resulting from an organisation's activities, products and services.

Structural technology is a synthetic issue, where multiple requirements are fulfilled through the application of several basic technologies such as materials technology, manufacturing technology, information and automation technology. Structural design includes creative phases, analysis phases as well as optimisation and synthesis phases. In these phases there are also many solution methods. In the creative phase several innovation and brainstorming methods can be applied. The analysis and optimisation phase requires such skills as applied mathematics, physics, chemistry and even biology. The same methods used in design can also be applied to structural product development.

Environmentally-efficient structural technology and design are the response to the goals defined in ISO 14004. The central problem in the development of sustainable structural technology and design is to recognise how the design process needs to be changed and which methods can be applied in order to realise the new kind of structural design which is needed.

Ecology can be linked to the environmental costs. The term gives us quite a concrete starting point for the application of this aspect in structural engineering and design. Sustainability is related not only to ecology and economy, but also to all other groups of requirements, and it can be applied in the life cycle methodology of design, manufacturing, construction and management. Through the principle of sustainability, resistance design will be expanded into durability design, thus introducing time as a new dimension in design calculations (Sarja and Vesikari 1996). Consequently aspects of maintenance, changes during use, modernisation, renewal, repair, demolition, recycling and disposal need also to be added. Health and safey aspects are generally related to the control of moisture and temperature conditions and to special subjects such as noxious emissions from materials. Structural systems can be divided into assemblies having quite different requirements and environmental properties. Buildings and engineering structures have many aspects similar to each other, but they also differ in some respects.

In buildings, energy consumption mostly dictates environmental properties. For this reason the thermal insulation of the envelope is important. The most important environmental aspects for structures with a long target service life are flexibility with regard to functional changes and high durability, while for structures with a moderate or short target service life changeability and recyclability are more important. The loadbearing frame is the most massive and longlasting part of the building. For the loadbearing frame, durability and flexibility to enable changes in the functions, spaces and services systems are most important. Internal walls have a moderate requirement of service life and may need to accommodate changes. For this reason internal walls must have good changeability and recyclability. An additional property of an environmentally effective structural system is good and flexible compatibility with the building services system, because the services system is the part of the building that is most often changed. During production it is important to apply effective recycling of production wastes in factories and on site. Finally there is the requirement to recycle the components and materials after demolition.

Engineering structures such as bridges, dams, towers and cooling towers are often very massive and their target service life is long. Therefore environmental efficiency depends on the selection of environmentally-friendly local raw materials, high durability and easy maintainability of the structures during use, recycling of construction wastes and finally recycling of the components and materials after demolition. Some parts of engineering structures, such as waterproofing membranes and railings, have a short or moderate service life and, consequently, easy re-assembly and recycling are most important.

1.4 Definition of integrated structural design

The objective of the development of environmentally-oriented design is to realise design methods and methodologies for structural design in order to meet the requirements of sustainable development during the entire life cycle of the buildings. Environmental design will be presented as a part of integrated life cycle design, which includes the mechanical, physical, economic, energy, health and environmental aspects.

The term 'integrated life cycle design' includes the design process, methodology and methods of design for life cycle quality, which aims to fulfil the multiple requirements of users, owners and society in an optimised way during the entire life cycle of a building or other built facility.

1.5 Content and schedule of integrated structural design

Structural design is an important link in construction, translating the requirements of owners, users and society into performance requirements of the technical systems, creating and optimising technical solutions, which fulfil those requirements, and proving through analysis and dimensioning calculations, that the performance requirements will be fulfilled over the entire design service life. The framework of integrated design is presented in Figure 1.1.

1.6 Methodologies of life cycle design

The integrated life cycle design methodology is aimed at regulating optimisation and guaranteeing the life cycle human conditions, economy, culture and ecology with technical performance parameters (Sarja 1994, 1995, 1996, 1997a, 1997b, 1997c, 1999a, 1999b, 2000a), as presented in Figure 1.2 (Sarja 2000b). With the aid of life cycle design we thus can optimise human conditions (health and safety, and comfort), and minimise financial costs and the environmental impacts. Through service life (performance and durability) design, the targeted service life can be guaranteed. The conceptual, creative design phase is decisive in using the potential benefits of integrated design process effectively. In this phase, the design is made on the system and module levels. Hierarchical modular systematics (Sarja 1989, Sarja and Hannus 1995) helps rational design, as the structural system typically has different parts or modules with different requirements, for example with regard to durability and service life requirements.

In life cycle design, analysis and design are expanded into two economic areas: financial and ecological (environmental costs and impacts). Life cycle costs are calculated as the present value or as yearly costs by discounting the costs from manufacture, construction, maintenance, repair, changes, modernisation, reuse, recycling and disposal. Monetary costs are treated as usual in current value calculations. Environmental costs are the use of non-renewable natural resources and the production of air, water or soil pollution. The consequence of air pollution includes health problems, inconvenience to people, ozone depletion and global warming. These impacts dictate the environmental profiles of structural and building services systems. The goal is to keep environmental costs below permitted values and to minimise them

The environmental impact profile generally includes the consumption of globally and locally critical raw materials such as energy and water and the

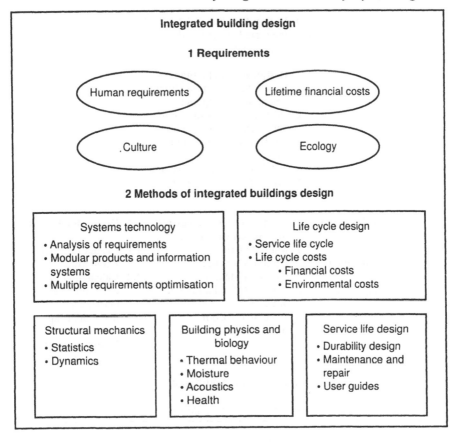

Figure 1.1 Framework of integrated building design

production of CO_2, CO, SO_2, NO_x, dust, solid wastes and noise. Environmental impact profiles are systematised in several approaches, such as ATHENA (Trusty and Paehlke 1994), the European Commission APAS-Programme report (Peuportier and REGENER Project 1997) and BREEAM methodology (BREEAM 1990). A comprehensive description on existing life cycle assessment methods is reported by the United States Environmental Protection Agency (US EPA 1995). The basic principles are treated at a general level in several reports (Holmberg 1995). It is also critical to introduce time factors into the detailed structural design. This includes service life design to guarantee that the durability and performance of the structures meets corresponding target values (Sarja and Vesikari 1996, Sarja 2000c).

Health impacts can be identified from medical knowledge and applied as criteria for structural design mainly in the areas of building physics and biology, including several hygrothermal and acoustical methods (Radünz 1998, Fava *et al.* 1992).

0 Design Process and Methods
0.1 Integrated structural design and
     its applications

2 Design for Performance
2.1 Functional design of structural
     system and components
2.2 Design for flexibility in use
     and in changes of the use
2.3 Hygrothermal design of
     structures
2.4 Detailed design of structures
2.5 Maintenance and repair
     design

3 Design for Economy	**1 Control of Requirements and Specifications**	**4 Design for Ecology**
3.1 Economic life cycle design of structural system and components	1.1 Multiple Criteria Optimisation and Decision Making 1.2 Service Life Planning of structures 1.3 Integrated design documentation	4.1 Design for Energy Economy 4.2 Calculation of environmental burdens in manufacturing and construction 4.3 Design for reuse and recycling

**5 Design for Health and
   Comfort**
5.1 Health aspects in design

Figure 1.2 Modules of the integrated life cycle design (Sarja 2000b)

1.7 Factors of sustainability

Generally speaking, the results of comparisons as regards ecological requirements, lead to the conclusion that differences between different materials and structural solutions during the construction phase are quite small. However, quite large differences can be found between life cycle sustainability factors of existing buildings or other facilities. These are caused by differences in the basic factors of sustainability: flexibility of design, buildability during the manufacturing and construction phase, adaptability to change during use, durability in comparison with the design service life and the recyclability of components with a quite short service life.

Energy consumption in buildings is economically important and it dictates most of the environmental properties of a building's life cycle, the differences in environmental costs between different structural systems is otherwise quite small. For example, in northern Europe the heating of buildings and provision of hot water and lighting produce, during a life cycle of 50 years, about 90 per cent of air pollutants, compared with only 10 per cent produced during the manufacturing of building materials, transport and construction. In warm and tropical countries the heating requirements are lower, but cooling buildings with air conditioning consumes even more energy than heating in cold climate countries. For this reason, as well as well-controlled heating, ventilating, cooling and heat recovery, the thermal insulation of the envelope is important. The loadbearing frame is the most massive and long-lasting part of the building, and its durability as well as its flexibility for future changes in use, layout and services provision are very important. The envelope needs to be durable and, as mentioned above, have effective thermal insulation and a safe static and hygrothermal behaviour. Internal walls have a more moderate service life, but they need to cope with relatively high degrees of change, and must therefore possess good adaptability and recyclability. An additional property of an environmentally-effective structural system is a good and flexible compatibility with the building services system, as the latter is the most frequently changed part of the building. In the production phase it is important to ensure the effective recycling of the production wastes in factories and on site.

With regard to transportation systems, the life cycle energy efficiency of the entire traffic system is also of utmost importance. Civil engineering structures like bridges, harbours, roads, railways etc. are often very massive and their target service life is long (Sarja 2000d). Their repair works under use are difficult. Their life cycle quality is tied to high durability and easy maintainability during use, saving of materials and the selection of environmentally-friendly raw materials, minimising and recycling of construction wastes, and finally recycling of the materials and components after demolition. Some parts of the engineering structures like waterproof membranes and bridge railings have a short or moderate service life and therefore the aspects of easy re-assembly and recycling are most important. Technical or performance-related obsolescence of the transport system and its structures is a major reason for the demolition of civil engineering structures, which raises the need for the careful planning of the whole civil engineering system. We know that cost is the decisive factor in our society and budgets must always stay within agreed limits. Cost plays a major role when decisions between design alternatives are being made.

1.8 Energy efficient building concepts and specifications

Heat energy consumption can be modelled as a function of the thermal insulation factors of the building envelope, interior ventilation and air leakage flow through the envelope. Thermal insulation factors include heat conduction through different parts of the envelope: walls, roof, floor, windows and doors. Ventilation includes the heat loss with exhaust air.

Typical parameters of multi-storey apartment buildings in a north European climate are presented in Table 1.1.

Table 1.1. Thermal parameters defining levels of energy efficiency in buildings

	Class of energy efficiency		
	Low energy efficiency	Normal energy efficiency	High energy efficiency (low energy building)
Walls: k (W/m² °K)	1.00	0.30	0.15
Roof: k (W/m² °K)	0.80	0.25	0.15
Ground floor: k (W/m² °K)	0.80	0.30	0.15
Windows: k (W/m² °K)	4.00	2.00	0.80
doors: k (W/m² °K)	3.00	0.70	0.30
$V_{air, vent}$ / hm² living floor area k (W/m² °K)	0.50	1.25	1.25
$V_{air, leak}$ / hm² living floor area k (W/m² °K)	1.50	0.50	0.10
heat recovery efficiency of ventilation, r	0.0	0.0	0.60

References

BREEAM (1990) *Environmental Assessment Method.* Building Research Establishment: Garston.

Fava, J., Roy, F., Consoli, F., Denison, R., Dickson, K., Mohin, T. and Vigon, B. (eds) (1992) *A Conceptual Framework for Life Cycle Impact Assessment.* Society of Environmental Toxicology and Chemistry and SETAC Foundation for Environmental Education, Inc.: Sandestin, FA.

Holmberg, J. (1995) 'Socio-ecological principles and indicators for sustainability'. Dissertation, Institute of Physical Resource Theory, Chalmers University of Technology, University of Göteborg.

ISO/DIS 15686-1. ISO TC 59/SC14. Guide for service life design of buildings. Draft standard.

Peuportier, B. (coordinator), Kohler, N., Boonstra, C., Blanc-Sommereux, I., Hamadou, H., Pagani, R., Gobin, C. and Kreider, J. (1997) *European Methodology for the Evaluation of Environmental Impact of Buildings: Life Cycle Assessment.* REGENER Project, Final Report to European Commission DG 12, Brussels.

Radünz, A. (1998) *Bauprodukte und gebäudebedingte Erkrankungen.* Springer Verlag: Berlin, Heidelberg.

Sarja, A. (1989) *Principles and Solutions of the New System Building Technology (TAT).* Research Report 662. Technical Research Centre of Finland: Espoo.

Sarja, A. (1994) 'Development towards the ecological and recyclable building materials technology'. Second International Conference on Materials Engineering for Resources, 19–22 October, Akita. Research Institute of Natural Resources attached to Mining College, Akita University: Akita.

Sarja, A. (1995) 'Methods and methodology for the environmental design of structures', RILEM Workshop on Environmental Aspects of Building Materials and Structures. Technical Research Centre of Finland: Espoo.

Sarja, A. (1996) 'Environmental design methods in materials and structural engineering'. *CIB Information,* 4(96), 23–25.

Sarja, A. (1997a) 'A vision of sustainable materials and structural engineering', in Tuutti, K. (ed.) *Selected Research Studies from Scandinavia.* Report TVBM-3078. Lund University, Lund Institute of Technology: Lund.

Sarja, A. (1997b) 'Framework and methods of life cycle design of buildings', Symposium: Recovery, Recycling, Reintegration, R'97, 4–7 February, Geneva. *EMPA*, VI, 100–105.

Sarja, A. (1997c) 'Some principles of integrated structural design', *Structural Engineering International (SEI)* 1/1997, 59–60.

Sarja, A. (1999a) 'Environmental design methods in materials and structural engineering', *Materials and Structures* 32, (December), 699–707.

Sarja, A. (1999b) 'Towards life cycle oriented structural engineering', in R. Eligehausen (ed.) *Construction Material: Theory and Application*. Ibiidem-Verlag: Stuttgart.

Sarja, A. (2000a) 'Integrated life cycle design as a key tool for sustainable construction', in Sarja, A. (ed.), *Integrated Life-Cycle Design of Materials and Structures ILCDES 2000*. RILEM Proceedings PRO 14. RIL – Association of Finnish Civil Engineers: Helsinki.

Sarja, A. (2000b) 'Development towards practical instructions of life cycle design in Finland', in Sarja, A. (ed.) *Integrated Life-Cycle Design of Materials and Structures ILCDES 2000*. RILEM Proceedings PRO 14. RIL – Association of Finnish Civil Engineers: Helsinki.

Sarja, A. (2000c) 'Durability design of concrete structures: Committee Report 130-CSL'. *Materials and Structures/Matériaux et Constructions*, 33 (January–February), 14–20.

Sarja, A. (2000d) 'Design of transportation structures for sustainability'. 16th Congress of IABSE, Lucerne, Congress Report, and CD ROM. IABSE-AIPC-IVBH: Zurich and ETH-Hönggenberg.

Sarja, A. (ed.) (2001) *Lifetime Structural Engineering* (in Finnish). Guidelines RIL 216-2001. Finnish Association of Civil Engineers: Helsinki.

Sarja, A. and Hannus, M. (1995) *Modular Systematics for the Industrialised Building*. VTT Publications 238. Technical Research Centre of Finland: Espoo.

Sarja, A. and Vesikari, E. (1996) *Durability Design of Concrete Structures*. RILEM Report Series 14. E&FN Spon: London.

Trusty, W.B. and Paehlke, R. (1994) *Assessing the Relative Ecological Carrying Capacity Impacts of Resource Extraction*. Forintek Canada Corp.: Vancouver.

US EPA (1995) *Life-Cycle Impact Assesssment: A Conceptual Framework, Key Issues, and Summary of Existing Methods*. Report EPA-452/R-95-002 United States Environmental Protection Agency: Washington, DC.

2 Design process

2.1 Design tasks, methodology and methods

Integrated structural life cycle design includes the tasks and methods in the design process shown in Table 2.1 (Sarja 1996, Sarja 2001).

2.2 Design process model

Integrated structural life cycle design includes the following main phases of the design process (see Table 2.2): investment planning, analysis of the actual requirements, translation of the requirements into technical performance specifications of structures, creation of alternative structural solutions, life cycle analysis and preliminary optimisation of the alternatives, selection of the optimal solution between the alternatives and the detailed design of the selected structural system. The conceptual, creative design phase is decisive if the potential benefits of integrated life cycle design process are to be utilised effectively. In this phase, the design is made at a system level. Modular systematics helps rational design, because the structural system typically has different parts (here called modules) with different requirements with regard, for example, to durability and service life requirements (Sarja and Hannus 1995).

The introduction of integrated principles into practical design is a wide-ranging process, where not only are the working methods of structural engineers changing, but also where co-operation between structural engineers, architects, building services system designers and other partners in construction and use has to be developed. Co-operation with clients and architects is especially important in order to utilise effectively the expertise of the structural engineers in the most decisive, creative and conceptual phases of the design. This type of co-operation also helps clients to realise the benefits of investing slightly more into structural design. Another important change in the design will be some kind of modularisation of the design, which means the separation of the designing of the functions, spaces and performance specifications from the designing of technical systems and modules. The first part of the design, which is performance oriented, will be made by the architect and technical designers in close co-operation with each other, as well as with the client and with users. The second part, which is a realisation

Table 2.1 Integrated life cycle design process and central methods for application

Design phase	Life cycle design methods
Investment planning	Multiple criteria analysis, optimisation and decision-making Life cycle (financial and environmental) costs
Analysis of clients' and users' needs	Modular design methodology Quality function deployment (QFD) method
Functional specifications of the buildings	Modular design methodology Quality function deployment (QFD) method
Technical performance specifications	Modular design methodology Quality function deployment (QFD) method
Creation and outlining of alternative structural solutions	Modular design methodology
Modular life cycle planning and service life optimisation of each alternative	Modular design methodology Modular service life planning Life cycle (financial and environmental) costs calculations
Multiple criteria ranking and selection between alternative solutions and products	Modular design methodology Quality function deployment (QFD) method Multiple criteria analysis, optimisation and decision-making
Detailed design of the selected solution	Design for future changes Design for durability Design for health Design for safety Design for hygrothermal performance Users' manual Design for re-use and recycling

phase, is carried out by teams of technical designers and manufacturers working closely together. The realisation phase is often connected to contractors' and supppliers' specific building concepts. In this way the current problem of divergent design and manufacturing processes can be avoided without compromising the functional, performance and other requirements for the life cycle use of the building, which were discussed earlier.

2.3 Design phases

These new aspects are widening the scope of the structural design and construction to the extent that the entire working process must be re-engineered. We can start to establish a new design process – integrated structural design – which will be described below.

Starting from investment planning and analysis of owners' and users' needs and requirements, the structural engineer must be ready to work with the architect

Table 2.2 Design phases, tasks and methods of integrated life cycle design

Design phase	Tasks (possible tasks for structural engineers is shown in **bold**)	Life cycle design methods
Investment planning	Define objectives of the building project.	Multiple criteria analysis, optimisation and decision-making.
	Define the study time period.	
	Create alternative investment plans.	
	Calculate life cycle (financial and environmental) costs (LCCs).	Life cycle financial and environmental costs.
	Calculate cash flows of alternative plans.	
	Evaluate benefits of the alternative plans.	
	Compare LCCs and make final decision.	
	Define final objectives.	
Analysis of clients' and users' needs	Identify relevant attributes (customers' requirements).	Modular design methodology.
	Estimate the rate of importance of each attribute as weight.	Quality function deployment (QFD) method.
Functional specifications of the buildings	Translate the results of needs analysis to demands.	Modular design methodology.
	Identify relevant functional properties.	Quality function deployment (QFD) method.
	Define weight of each property.	
Technical performance specifications	Translate functional properties and their weights from previous task to demands.	Modular design methodology.
	Identify technical performance properties.	Quality function deployment (QFD) method.
	Identify weight of each property.	
Creation and outlining of alternative structural solutions	Create and outline alternative solutions for building, its structural systems and building services in co-operation with other designers and project partners.	Modular design methodology.

(continued...)

Table 2.2 (continued)

Design phase	Tasks	Life cycle design methods
Modular life cycle planning and service life optimisation of each alternative	Define the requirement for design service life of the building.	Modular design methodology.
	Divide the building into service life modules with different service life classes.	Modular service life planning.
	Identify the number of life cycles of each module during the design service life of the building.	Life cycle (financial and environmental) costs calculations.
	Identify the design life cycle costs (monetary and enviromental) of the modules.	
	Outline alternative service lifes for the modules.	
	Define optimal service lifes for the modules, based on minimum total costs (financial and environmental).	
Multiple criteria ranking and selection between alternative solutions and products	Transfer the optimised service life cost of each alternative building concept from previous tasks.	Modular design methodology.
	Define multiple attributes from analysis of owners' and users' requirements.	Quality function deployment (QFD) method.
	Evaluate the performance properties of each alternative.	Multiple criteria analysis, optimisation and decision-making.
	Select the alternative for realisation between the alternatives.	
Detailed design of the selected solution	Design the structural modules for different performance requirements.	Design for future changes.
		Design for durability.
	Make the synthetic design.	Design for health.
		Design for safety.
		Design for hygrothermal performance.
		Users' manual.
		Design for reuse and recycling.

and other partners in the building process. Controlled and rational decision-making when optimising between the many requirements with different measurement systems is possible through the use of multiple criteria decision-making. Environmental and health and safety aspects are are given more weight. Service life principles introduce time as a variable in economics and design.

2.3.1 Investment planning

The owner/client defines life cycle objectives such as area and functional requirements of building and its spaces, economics, requirements for use, service life, aesthetic objectives and ecological objectives.

Designers in co-operation with the owner create alternative investment plans, and make a multiple criteria analysis leading to decision-making between alternatives.

2.3.2 Analysis of clients' and users' needs

The analysis of clients' needs is a preparatory phase of the building project. This phase deals mainly with the use and the spatial requirements of the buildings, and it is the responsibility of the client, the architect and the contractor. Usually the structural designer is not involved in this analysis, but provides technical support for the architect and uses the results as a starting point for his own work.

2.3.3 Translation of the clients' and users' needs into functional life cycle requirements of the buildings

The quality function deployment (QFD) method can be applied as a means of analysis when ranking the functional properties of draft design alternatives in comparison to the owners' and users' needs. The architect has main responsibility of this phase, and the structural engineer provides technical support and expertise and is possibly involved in calculations connected with the analysis.

2.3.4 Translation of the functional life cycle requirements into technical performance specifications

The results of the functional analysis and planning will be converted into structural requirements in co-operation between the architect and the structural designer. The result is a set of technical specifications, including mechanical, physical, aesthetic, environmental, energy and health specifications. The performance specifications are presented in detail in ISO 6240-1980: 'Performance standards in building – Contents and presentation' and ISO 6241-1984: 'Performance standards in building – Principles for their preparation and factors to be considered'.

Once the clients' needs have been properly defined and translated into functional requirements, they must be expressed as performance specifications. They should, as far as possible, be quantitative and preferably related to well-established (standardised) test methods. However, these methods are only available to a limited extent and consequently performance requirements must be given in other forms. Often the requirements are described only in qualitative terms. Requirements or 'target figures' for heating and cooling energy needs are given in some cases, as well as financial figures such as annual costs for the building or its sub-systems.

Modular scheduling and allocation at the conceptual design phase includes the specification of the alternative structural solutions with regard to the target service life and technical performance requirements of each structural module. Based on the specifications, estimates of lifetime financial and environmental costs as current values or annual costs are calculated. The model for the modular specification of the technical performance properties is presented in Table 2.3. The specification work must interact with life cycle optimisation of the central properties in order to reach the target of optimal design (ASTM 1995, CEN 1997).

2.3.5 Creation and outlining of alternative structural solutions

The phase of conceptual design is the most important, because the decisions made in this phase have a strong influence on the economic, environmental, aesthetic, functional and technical quality of structures. The role of structural designer in this phase is to support decision-making by the owner and users. The main tools in this work are creative structural drafting, analysis, optimisation, and multiple criteria decision-making described above. The role of the architect is very similar to that of the structural designer, but the tools are more weighted towards creative functional and aesthetic areas. Close co-operation between the structural designer, the architect and building services designers is needed.

Living and working demands on spaces and other functional requirements of buildings are changing more and more rapidly. The future value of buildings is largely dependent on their flexibility for changes in use. The structural system and its compatibility with the building services system are decisive factors for adaptability of a building. Starting as early as the sketch design, design for change is one of the central aims. Important issues in conceptual design are:

- sufficient storey height to allow for changes in services installations
- long spans of floorings which do not prevent changes in use because of too many vertical loadbearing structures
- location of openings in facades
- spaces for staircases (connection modules) for traffic, and for horizontal and vertical connections of building services systems
- space for horizontal piping and wiring
- room sizes large enough for alternative uses
- easily moveable and reusable partition walls
- easily changed electrical and communication wiring systems.

Table 2.3 Specification of performance properties for alternative structural solutions at a module level using a multi-storey apartment building as an example

Structural module	Central performance properties in specifications
Foundations	Loadbearing capacity, target service life, limits and targets for environmental impact profiles.
Loadbearing frame	Loadbearing capacity, target service life, estimated repair intervals, estimated maintenance costs, limits and targets for environmental impact profiles.
Envelope/walls	Target values for thermal insulation, target service life, estimated repair intervals, estimated maintenance costs, limits and targets for environmental impact profiles.
Envelope/roof	Target values for thermal insulation, target service life, estimated repair intervals, estimated maintenance costs, limits and targets for environmental impact profiles.
Envelope/ground floor	Target values for thermal insulation, target service life, estimated repair intervals, estimated maintenance costs, limits and targets of environmental impact profiles.
Envelope/windows	Target values for thermal insulation, target service life, estimated repair intervals, estimated maintenance costs, limits and targets for environmental impact profiles.
Envelope/doors	Target values for thermal insulation, target service life, estimated repair intervals, estimated maintenance costs, limits and targets for environmental impact profiles.
Partition floors	Target values for sound insulation, target service life, estimated repair intervals, estimated maintenance costs, limits and targets for environmental impact profiles, estimated intervals of the renewal of connected building service installations.
Partition walls (including doors)	Target values for sound insulation, target service life, estimated intervals of spatial changes in the building, estimated repair intervals, estimated maintenance costs, limits and targets for environmental impact profiles, estimated intervals of the renewal of connected building service installations.
Bathrooms and kitchens	Target values for sound and moisture insulation, target service life, estimated repair intervals, estimated maintenance costs, limits and targets of environmental impact profiles, estimated intervals of the renewal of connected building service installations.

2.3.6 Modular life cycle planning and service life optimisation

For life cycle planning, a modular methodology is preferred. This allows the systematic allocation and optimisation of the target service life as well as life cycle economy and ecology of different parts of the building (ASTM 1995, CEN 1997). A suitable modularisation at the highest level of hierarchy is as follows: loadbearing frame, envelope, foundations, partitions, heating and ventilating services, information, water and sewage systems, control services and waste management systems. All of these systems are specified during the development or design process with continuously increasing precision starting from general performance specifications and ending with detailed designs.

The following tasks are required for each design alternative:

- classification of building modules into target service life classes, following a suitable classification system.;
- defining the number of times each module must be renewed or replaced during the design service life of the building;
- calculation of total life cycle financial and environmental costs during the design life cycle of the building;
- preliminary optimisation of the total life cycle cost by varying the value of service life of key modules in each alternative within permitted values.

The division of the building into modules can be the same as that presented in Section 2.3.5. After the life cycle financial and environmental costs of alternative designs have been calculated, they are transferred into multiple criteria ranking and selection between alternative designs made.

2.3.7 Ranking and selection between design alternatives and products

The ranking of design alternatives ends the sketch design phase and results in the draft designs. Even in the sketch design phase, the selection of some key products which are connected to solutions of the structural system may have been partly completed; although the selection of products is mainly done at later phases of the design. When applying integrated design procedures, all classes of requirements are systematically taken into account during the ranking. Multi-attribute decision analysis (MADA) is used as a ranking method (ASTM 1995, Roozenburg, and Eekels 1990, Sarja 1999a, Sarja 2001). Core properties are principally calculated quantitatively with numerical values, but some additional properties are only evaluated qualitatively. The properties are normalised by comparison with a reference alternative. This phase of design is usually the responsibility of the architect who is supported by the structural designer and the designers of building services systems. An example of a multiple-attribute decision- making procedure in design is presented in Figure 2.1.

2.3.8 Detailed design of the selected solution

The role of detailed design phase is to ensure that the targets and specifications defined in the conceptual design can be realised in construction and throughout the building's life cycle. This means the structure must be buildable, serviceable, durable, that it can be maintained and repaired, and finally that it can be demolished and the waste recycled or disposed of. The methods used at this phase include:

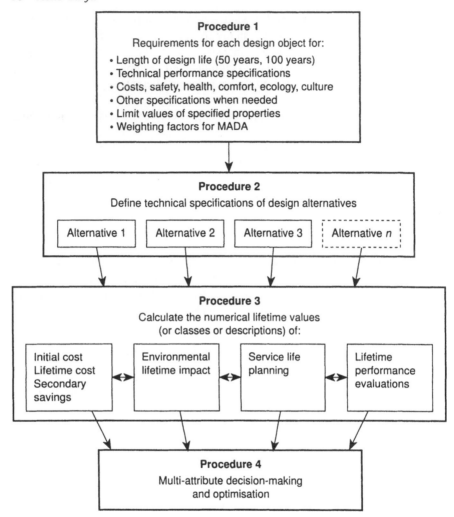

Figure 2.1 Multiple-attribute decision-making procedure

structural mechanics, building physics (moisture and thermal calculation methods, and methods of acoustics and fire resistance) and durability design.

The detailed design includes generally the following phases:

1 ordinary mechanical design
2 durability design
3 final design.

Ordinary mechanical design is performed using conventional design methods. Its purpose is to determine the preliminary dimensions for the structure.

The durability design procedure is different for structures made from different materials. With concrete structures, most often a basic procedure can be applied. Usually the durability design procedure for steel structures also follows a general procedure. The durability of wooden structures is connected to moisture and temperature and thus leads to the moisture control of structures in order to eliminate the danger of the wood rotting. When using deterioration calculation models, the design procedure for wooden structures also follows the general procedure.

A flow chart of the design procedure, with the design of concrete structures as an example, is presented in Figure 2.2

2.3.8.1 Static, dynamic and seismic design

Mechanical design includes the static, fatigue and dynamic design aspects. This design is traditional and many manuals, guides, norms and standards exist, therefore this phase will not be discussed any further in this context.

2.3.8.2 Durability design for service life

The role of durability design is important in life cycle design. The objective of durability design is to ensure that the specified target service life can be achieved in the actual working environment of the structure.

In ordinary design the durability for ordinary service life, generally for 50 years, is taken into account through the structural detailing rules found in norms, standards and design manuals. When using a target service life other than 50 years separate service life design calculations are needed. For these specific purposes statistically-based service life design methods can be used to produce specific detailing rules and model designs which then can be applied for similar specific cases. Statistically-based life cycle design can also be applied in the product development of prefabricated structural units.

The detailed durability design procedure is as follows (Sarja and Vesikari 1996, Sarja 2000b, Sarja 1997b):

1 specification of the target service life and design service life
2 analysis of environmental effects
3 identification of durability factors and degradation mechanisms
4 selection of a durability calculation model for each degradation mechanism
5 calculation of durability parameters using available calculation models
6 possible updating of the calculations of the ordinary mechanical design (e.g. own weight of structures)
7 transfer of the durability parameters into the final design.

LIFETIME SAFETY FACTOR METHOD

The lifetime safety factor design procedure is somewhat different for structures consisting of different materials, although the basic design procedure is the same

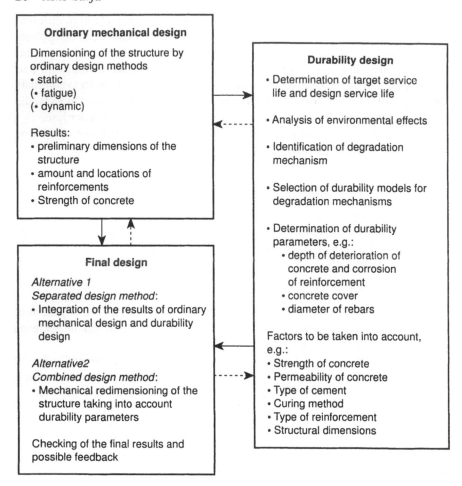

Figure 2.2 Flow chart of the durability design procedure, concrete structures as an example (Sarja and Vesikari 1996, Sarja 2000a)

for all kinds of materials and structures. The lifetime safety factor method is analogous with the safety factor method of static design, also known as limit state design.

The lifetime factor design procedure is as follows:

1   specification of target service life and design service life
2   analysis of environmental effects
3   identification of durability factors and degradation mechanisms
4   selection of a durability calculation model for each degradation mechanism
5   calculation of durability parameters using available calculation models
6   possible updating of calculations of the ordinary mechanical design (e.g. own weight of structures)
7   transfer of durability parameters into final design.

STATISTICAL DURABILITY DESIGN

The simplest mathematical model for describing the 'failure' event comprises a load variable S and a response variable R (Sarja and Vesikari 1996). In principle the variables S and R can be any quantities and expressed in any units. The only requirement is that they are commensurable. Thus, for example, S can be a weathering effect and can be the capability of the surface to resist the weathering effect without unacceptably large visual damage or loss of the reinforcement concrete cover.

If R and S are independent of time, the 'failure' event can be expressed as follows (Sarja and Vesikari 1996):

$$\{\text{failure}\} = \{R < S\} \tag{2.3}$$

The failure probability P_f is now defined as the probability of that 'failure':

$$P_f = P\{R < S\} \tag{2.4}$$

Either the resistance R or the load S or both can be time-dependent quantities. Thus the failure probability is also a time-dependent quantity. Considering $R(\tau)$ and $S(\tau)$ are instantaneous physical values of the resistance and the load at the moment τ the failure probability in a lifetime t could be defined as:

$$P_f(t) = P\{R(\tau) < S(\tau)\} \text{ for all } \tau \le t \tag{2.5a}$$

The determination of the function $P_f(t)$ according to the Equation 2.5a is mathematically difficult. That is why R and S are considered to be stochastic quantities with time-dependent or constant density distributions. By this means the failure probability can usually be defined as:

$$P_f(t) = P\{R(t) < S(t)\} \tag{2.5b}$$

According to the Equation 2.5b the failure probability increases continuously with time as schematically presented in Figure 2.3.

Considering continuous distributions, the failure probability P_f at a certain moment of time can be determined using the convolution integral:

$$P_f = \int F_R(s) f_s(s) \mathrm{d}s \tag{2.6}$$

where $F_R(s)$ is   the distribution function of R,
   $f_s(s)$ the probability density function of S, and
   s the common quantity or measure of R and S.

The integral can be approximately solved by numerical methods.

The statistical method can, in principle, be used for individual special cases even in practice, but this will not find any common use. The main use of statistical

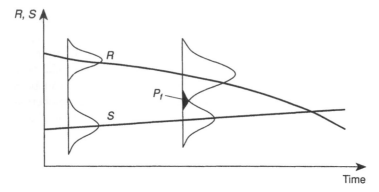

Figure 2.3 The increase of failure probability: illustrative presentation
(Sarja and Vesikari 1996)

theory is in the development of deterministic methods. Such a method is the lifetime
safety factor method presented above.

2.3.8.3 Hygrothermal design

In many countries, modern building has demonstrated significant problems
associated with the indoor climate. The expression 'sick building syndrome' (SBS)
is used to characterise the problem. Much research has been performed in order to
clarify the mechanisms, but the final answer has not yet been found. Drastically
increased thermal insulation and airtightness of buildings following the first so-
called oil crisis in 1973, when energy prices increased rapidly, is often blamed for
the problems. However, it seems obvious that moisture content and transport in
materials and in building components play an important role. The well-known
problems in connection with foundation systems using concrete slabs directly on
the ground and crawl space foundations could exemplify this.

As moisture dynamics are closely related to, and very often impossible to
separate from, thermal mechanisms it is obvious that the hygrothermal analysis
of the building should be undertaken. Design methods are available today for the
most important performance requirements, but it must be stated that they are not
accurate as those used in strength and deformation analysis. One reason for this is
that the measurement of material properties is, in many respects, much more
difficult, especially concerning moisture transport. Much of the basic physics has
not been fully clarified.

Important material properties in the hygrothermal analysis are:

*   heat conductivity
*   specific heat
*   thermal diffusivity
*   vapour resistance
*   moisture diffusivity

- capillarity
- sorption curves (relative humidity versus moisture content at equilibrium).

The hygrothermal design of a building contains the following major steps:

- definition of building context
- clarification of thermal and moisture environment (boundary conditions)
- analysis of temperatures and moisture status in different parts of the structure, account being taken to non-steady state conditions.
- definition of critical temperature and moisture levels
- comparison of actual and critical values
- accepting/rejecting the proposed solution.

2.3.8.4 Acoustic design

Acoustic design usually includes airborne sound insulation, impact noise level control between spaces and control of vibrations of structures during use. Sound insulation is usually controlled through standards on sound insulation and vibrations, but special rules and calculation methods (e.g. for flooring) also exist.

In special cases (e.g. concert halls and theatres), the internal sound acoustics are a primary requirement. This is a job for specialised experts and is not dealt with in detail here.

2.3.8.5 Design for safety

Static and dynamic safety as well as fire safety are defined in international and national regulations, norms and standards dealing with traditional methods of mechanical design. Because these design methods are standard, they are not discussed here. Integrated life cycle design provides a new viewpoint on safety by a systematic consideration and optimisation of long-term safety, taking into account degradation effects.

2.3.8.6 Design for health

Health requirements can follow the guidelines of SETAC (Fava *et al.* 1992) and other national and international codes, standards and guides. The main issues are to avoid moisture in structures and on finishing surfaces, and to check that no materials used cause emissions or radiation which are dangerous for health and comfort of the users (Radünz 1998). In some areas radiation from the ground must be also be eliminated though insulation and ventilation of the foundations. Thus the main tools for health design are: selection of materials (especially finishing materials), eliminating risks of moisture in structures (through waterproofing, drying during construction and ventilation), and elimination of possible radioactive ground radiation with airproofing and ventilation of ground structures (Sarja 2001, Sarja 2000a).

2.3.8.7 Design for reuse and recycling

The environmental profile of basic materials already includes recycling efficiency, which means the environmental cost of recycling. It is important to recognise that the recycling potential of building components, modules and even technical systems need to reconsidered in connection with design. The higher the hierarchical level of recycling, the higher also the ecological and economic efficiency of recycling (Lippiatt 1998, Sarja 1999).

Designers can influence reuseability and recyclability through the choice of the structural system, component types and their connections, and through the choice of materials. Modules and components can be divided into different classes of reuseability and recycleability through performance and service life based modular systematics. Typically modules and components with a short service life or high likelihood of change should have a good potential for reuse or recycling. Modules and components which have a long service life, typically the loadbearing frame, must allow the layout of building spaces to be easily changed, and they must have a long service life and good maintainability.

Special issues to be considered in the design of structures and materials for reuse and recycling are:

- modular structural system with autonomous modules with a dimensional and modulation and clear tolerance system (Sarja 1989, Sarja and Hannus 1995)
- separability of the structural components or materials during demolition of building (e.g. through the use of demountable structural components using suitable connections and joints)
- constructive separation of technical systems (e.g structures and building services systems) with different service lives and different recycling techniques
- reduction in the variety of materials
- ability to separate materials, which cannot be recycled together
- avoiding insoluble composite substances and/or composite substances that are either only slightly soluble or soluble only within a high expenditure or energy input.

2.3.8.8 Users' manual and maintenance plan

A building, like a car or other equipment, needs a users' manual. The manual will be produced gradually during the design process in co-operation between the partners in design, manufacture and construction. The usual tasks of the structural designer include:

- collecting lists of maintenance tasks for the structural system
- collecting and applying instructions for operation, control and maintenance procedures and works
- checking and co-ordination of operation, control and maintenance instructions of product suppliers and of the contractor

- preparing the relevant parts for the users' manual
- checking the relevant parts of the final users' manual.

The main information sources for the users' manual are product descriptions from producers, which have to include relevant information on all classes of technical performance, service life, maintenance tasks and their frequency, and health aspects. Information on the environmental impact profile, resue and recycling should also be included in the product description.

The users' manual is a link between design and maintenance, thus paving the way towards real life time engineering (Sarja 1999b).

2.2.8.9 Final integration of design specifications

Design specifications which are produced at different phases of the design process are collected together during the design process, co-ordinated and checked to avoid differences in details and specifications. This co-ordination is a continuing task during the design period and it finally produces the documentation for building control and maintenance during the building's life cycle.

References

ASTM (1995) *Standard Practice for Applying the Analytic Hierarchy Process to Multi-attribute Decision Analysis of Investments Related to Buildings and Building Systems.* ASTM Designation E 1765-95.

CEN (1997) *Concrete: Performance, Production and Conformity.* Draft prEN 202. European Committee for Standardisation (CEN): Brussels.

Fava, J., Roy, F., Consoli, F., Denison, R., Dickson, K., Mohin, T. and Vigon, B. (eds) (1992) *A Conceptual Framework for Life Cycle Impact Assessment.* Society of Environmental Toxicology and Chemistry and SETAC Foundation for Environmental Education, Inc.: Sandestin, FA.

Lippiatt, B. (1998) 'Building for Environmental and Economic Sustainability (BEES)'. CIB/RILEM Symposium: Materials and Technologies for Sustainable Construction, Gävle, June. Building and Fire Research Laboratory, National Institute of Standards and Technology: Gathersburg, MD.

Radünz, A. (1998) *Bauprodukte und gebäudebedingte Erkrankungen.* Springer Verlag: Berlin, Heidelberg.

Roozenburg, N. and Eekels, J. (1990) *EVAD, Evaluation and Decision in Design. (Bewerten und Entscheiden beim Konstruiren).* Schriftenreihe WDK 17, Edition HEURISTA: Zürich.

Sarja, A. (1989) *Principles and Solutions of the New System Building Technology (TAT).* Research Report 662. Technical Research Centre of Finland: Espoo.

Sarja, A. (1996) 'Environmental design methods in materials and structural engineering'. *CIB Information*, 4(96), 23–25.

Sarja, A. (1997b) 'Framework and methods of life cycle design of buildings'. Symposium: Recovery, Recycling, Reintegration, R'97, 4–7 February, Geneva. EMPA, VI, 100–105.

Sarja, A. (1999a) *Environmental Design Methods in Materials and Mtructural Engineering* – Progress Report of RILEM TC 172-EDM / CIB TG 22. Materials and Structures, Vol. 32, December 1999, 699–707.

Sarja, A. (1999b) 'Towards life cycle oriented structural engineering', in Eligehausen, R. (ed.) *Construction Materials: Theory and Application.* Ibiidem-Verlag: Stuttgart.

Sarja, A. (2000a) 'Development towards practical instructions of life cycle design in Finland', in Sarja, A. (ed.) *Integrated Life-Cycle Design of Materials and Structures ILCDES 2000.* RILEM Proceedings PRO 14. RIL – Association of Finnish Civil Engineers: Helsinki.

Sarja, A. (2000b) 'Durability design of concrete structures: Committee Report 130-CSL'. *Materials and Structures/Matériaux et Constructions*, 33 (January–February), 14–20.

Sarja, A. (2001) *Lifetime Structural Engineering* (in Finnish). Guidelines RIL 216-2001. Finnish Association of Civil Engineers: Helsinki.

Sarja, A. and Hannus, M. (1995) *Modular Systematics for the Industrialised Building.* VTT Publications 238, Technical Research Centre of Finland: Espoo.

Sarja, A. and Vesikari, E. (1996) *Durability Design of Concrete Structures.* RILEM Report Series 14. E&FN Spon: London.

3 Life cycle design methods

3.1 Classification of design methods

Life cycle quality as defined earlier in the text and Figure 0.1 in the Introduction, can be implemented in design with different types of methods and principles. A classification of these methods is presented in Figure 3.1.

Life cycle design methods are tools to guarantee life cycle quality in design. The design methods can be classified in relation to the factors of life cycle quality as presented in Table 3.1. The methods shown in Figure 3.1 and Table 3.1 are mainly applied to design of buildings.

The conceptual, creative design phase is decisive in ensuring that the potential benefits of the integrated design process are utilised effectively. These include traffic system planning and long-term optimisation, starting from regional planning and the planning of urban areas. Controlled and rational decision-making when optimising multiple requirements with evaluation criteria is possible through the application of systematic multiple attribute optimisation and decision-making. In the detailed design phase, life cycle aspects emphasise the need for total performance over the life cycle, including durability design and design for mechanical and hygrothermal performance.

The incorporation of integrated design principles into practical design is a fairly extensive process, in which not only is the work of the structural engineers changing, but also co-operation has to be developed between structural engineers and other partners in construction and use. A modularisation of the design – the separation of the functional design and performance specifications from the detailed design of structural systems and modules – will be needed. The first part of the design process is performance-oriented. The second part is created by a close team of technical designers and contractors. In this way the current problem of diversified design and manufacturing processes can be avoided without compromising functional and performance requirements.

This book focuses on those methods which are non-traditional and which fall within structural engineers' areas of expertise. Consequently, design of heating, ventilating, cooling and lighting systems as well as aesthetic design are not dealt with here, as they fall outside these areas of expertise. The usual areas of structural design: static, dynamic, seismic, hygrothermal and fire safety design have been mentioned only briefly, because they are a traditional area of structural design.

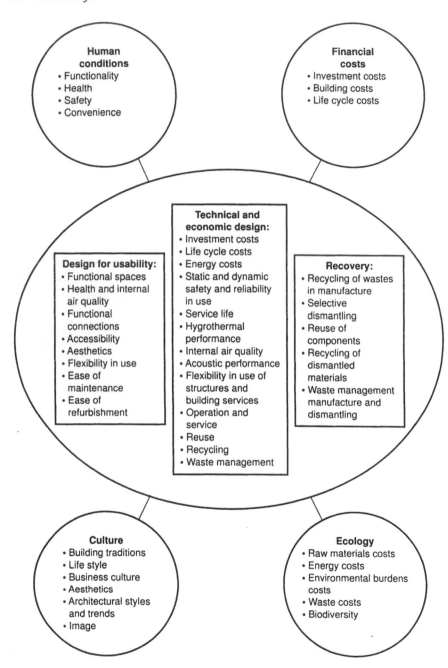

Figure 3.1 Classification of design methods of integrated life cycle design

Table 3.1 Design methods for different types of requirements of buildings

Requirement class	Design method
Functionality in use	Analysis of clients' needs. Multi-attribute analysis and decision-making. Quality function deployment method (QFD). Modular functional spatial planning. Design for changes in use. Modular service life planning and optimisation.
Financial considerations	Life cycle costing. Modular design methodology. Multiple-attribute optimisation and decision-making Modular service life planning and optimisation Service life design
Environmental considerations	Environmental life cycle analysis and costing. Modular design methodology. Design for energy efficiency. Design for recycling and reuse. Multi-attribute optimisation and decision-making.
Safety	Static design. Dynamic design. Service life and durability design. Seismic design. Fire safety design.
Health	Hygrothermal design. Health checking of materials and materials combinations. Health classification of internal climate. Design of heating, ventilation, cooling and lighting. systems for healthy interior conditions.
Convenience	Design for acoustic comfort. Design for comfortable internal climate. Design for aesthetic comfort.
Integration of design solutions	Aggregation of life cycle performance parameters. Multi-attribute analysis and decision-making.

In the design process of civil engineering structures some special viewpoints have to be taken into account. The main phases of the life cycle design procedure of a transportation system are: regional planning, planning of traffic and transport systems, investment planning, analysis of the actual requirements, translation of the requirements into technical performance specifications for structures, the creation of alternative structural solutions, life cycle analysis and preliminary optimisation of the alternatives, selection of the optimal solution between the alternatives, and finally the detailed design of the selected system and its structures. A summary of the integrated life cycle design phases and the specific design methods are presented in Table 3.2.

Table 3.2 Integrated life cycle design process and central methods for application in civil
         engineering

Design phase	Life cycle design methods
Regional planning, planning traffic and transport systems, investment planning	Multiple criteria analysis, optimisation and decision-making. Life cycle financial and environmental costs .
Analysis of users' and society's requirements	Multiple criteria analysis, optimisation and decision-making. Quality function deployment method (QFD).
Functional specifications and service life planning	Multiple criteria analysis, optimisation and decision-making. Quality function deployment method (QFD).
Technical performance specifications	Performance-based methodology.
Creation and outlining of alternative solutions	Modular design methodology.
Service life planning and service life optimisation	Modular design methodology. Modular service life planning. Life cycle financial and environmental cost calculations.
Multiple criteria ranking and selection between alternative solutions	Modular design methodology. Multiple criteria analysis, optimisation and decision-making.
Detailed design of the selected solution	Design for future changes. Design for durability. Design for safety. Design for multiple requirements performance. Design for health and comfort. User's manual. Design for re-use and recycling.

3.2 Modular design methodology

Open modular systematics includes modularisation of products, organisation and information, dimensional co-ordination, tolerance system, performance-based product specifications, product data models etc., so that the suppliers provide products and service modules that will fit together.

Openness is a concept with many aspects, for example

- opennness to competition between suppliers
- openness for alternative assemblies
- openness to future changes
- openness to information exchange
- opennnes for integration of modules and subsystems.

At all phases of the life cycle of a building, the hierarchical modular product systematics of the building can be applied (see Figure 3.2). Modular systematics

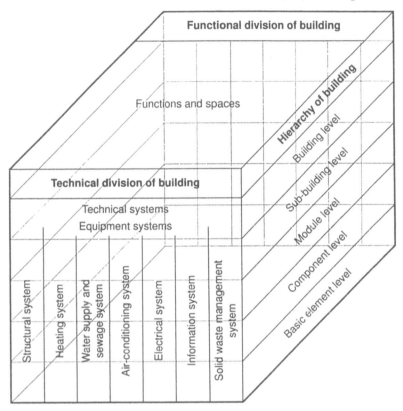

Figure 3.2 Overview of the hierarchical modular building product system (Sarja 1989, Sarja and Hannus 1995)

can be used to divide different parts of the building into different classes of service life. Typically the loadbearing frame represents a long-term service life, which has to be flexible to changes in the use and space allocation in the building. The floors also serve as the horizontal distributors of building services systems, and compatibility between the floors and the building's installations is extremely important. The overall product system aims to create an entire building from interacting parts. The system can thus be defined as an organised whole consisting of parts, the relations between which are defined by rules.

Compatibility can be achieved either through flexible integration or through separation of the structure and its installations. The envelope usually also has a requirement of a long service life and of good maintainability and repairability. Partition walls need to be altered when functional spaces are changed. In addition, partition walls are important locations of building services systems. The most concentrated service distribution parts of the building are the connection modules, which include the staircase, lift, vertical piping and wiring, horizontal distributing connections for service systems and possibly distributed building services equipment. It is important to develop building services systems (heating, ventilating, cooling, water and sewage, electrical, information and communication, and waste management) according to modular principles, taking into account especially the interaction and compatibility with the structural system. Building services functions can be distributed by the technical services systems in innovative ways, for example, by combining heating, cooling and ventilating together with a computerised operational control system into an integrated module. The systematics will be presented as model designs, alternative organisational models and applied product data models. It is important to identify and analyse productivity so that the results can then be used in the development of methods to improve productivity.

In modular design, modularisation involves division of the whole into sub-entities, which are to a significant extent compatible but independent. Compatibility makes it possible to use interchangeable products and designs which can be joined together according to connection rules to form a functional whole building or other structural system.

Typical modules in a building are:

- loadbearing frame
- facades
- roofing system
- partition walls
- building services systems.

Typical modules of a bridge are:

- foundations (including pilings)
- supporting vertical structures
- bearing horizontal structure
- deck
- deck waterproofing
- pavement
- edge beams
- railings.

Modular product systematics is connected to the performance systematics of the building. For example, the main performance requirements of floors can be classified in the following way:

1 Mechanical requirements, including

- static loadbearing capacity
- serviceability behaviour: deflection limits, cracking limits and damping of vibrations.

2 Physical requirements, including

- air tightness
- acoustics: airborne sound insulation, impact sound insulation, emission

- moisture tightness (in wet parts of the floor)
- thermal insulation between cold and warm spaces
- fire resistance and fire insulation.

3 Flexible compatibility with connecting structures and installations.

- partitions
- services: piping, wiring, heating and ventilating installations.

4 Other requirements:

- buildability
- changeability during the use
- easy demolition, reuse, recycling and wasting.

3.2 Life cycle costs

3.2.1 Financial and environmental costs

Future costs are calculated by estimating annual costs, recurring costs and their frequency and finally discounting future costs into current values.

Annual and recurring costs can be estimated simply as an addition of the relevant costs: construction costs, operation costs, maintenance costs, change costs, and repair, restoration and renewal costs. The costs of demolition, reuse, recycling and waste disposal can even be included. All these costs are discounted into current value, or calculated as annual costs.

As an example, the estimation of the future design service life costs of the alternative structural solutions of bridges is based on the specifications given in Table 3.3. In order to calculate life cycle costs, the frequency of maintenance, changes, repairs, restoration and renewal must be known.

Table 3.3 Specification of performance properties for the alternative structural solutions on a module level; as an example a bridge

Structural assembly	Central performance properties in the specifications
Substructures (foundations, retaining walls)	Loadbearing capacity, target service life, estimated repair intervals, estimated maintenance costs, limits and targets for environmental impact profiles
Vertical and horizontal superstructures (loadbearing structural system)	Loadbearing capacity, target service life, estimated repair intervals, estimated maintenance costs, limits and targets for environmental impact profiles
Deck overlayers (waterproofing, pavement)	Target values for moisture insulation, target service life, estimated repair intervals, estimated maintenance costs, limits and targets for environmental impact profiles, estimated intervals for renewal
Installations (railings, lights)	Target service life, estimated repair intervals, estimated maintenance costs, limits and targets for environmental impact profiles, estimated intervals for renewal

3.2.2 Calculating present values of life cycle financial and environmental costs

The building incurs financial and environmental costs during its entire life cycle starting from the planning phase and ending at recycling or waste disposal after demolition. The cost process, from which both monetary and environmental costing calculations can can be produced, is shown in Figure 3.3.

Life cycle costs are calculated using the usual current value discounting method, which is applied to both the financial and environmental costs, using the following equations

$$E_{tot}(t_d) = E(0) + \Sigma[N(t) \times E(t)] - E_r(t), \tag{3.1}$$

where

E_{tot} is   the design life cycle monetary cost as a present value
t_d design life
$E(0)$ construction cost
$N(t)$ coefficient for calculation of the current value of the cost at the time t after construction
$E(t)$ cost to be borne at time t after construction.
$E_r(t)$ residual value at time t

$$Ee_{tot}(t_d) = Ee(0) + \Sigma[N(t) \times (1-k_r) \times E_e(t)] \tag{3.2}$$

where

$Ee_{tot}(t_d)$ is   total life cycle environmental cost in relevant terms
$Ee(0)$ environmental cost at the construction phase
$Ee(t)$ environmental cost to be borne at time t after construction
k_r efficiency factor of recycling at renewal of each product

The time coefficient $N(t)$ can be calculated using the equations

$$N(t) = 1/(1+i)^n \tag{3.3}$$

where

i is   the interest rate
n the time (years) from the date of discounting

For the ecological calculations the expenses are environmental burdens, for example, consumption of non-renewable raw materials and energy, and the production of pollutants into the air, soil and water, including CO_2, CO, SO_2,

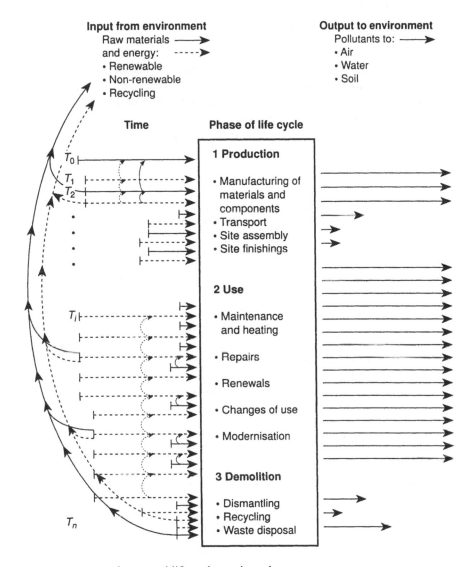

Input from environment
Raw materials ⟶
and energy: ---->
• Renewable
• Non-renewable
• Recycling

Output to environment
Pollutants to: ⟶
• Air
• Water
• Soil

Time   **Phase of life cycle**

T_0

1 Production

T_1
T_2

• Manufacturing of
  materials and
  components
• Transport
• Site assembly
• Site finishings

2 Use

T_i

• Maintenance
  and heating

• Repairs

• Renewals

• Changes of use

• Modernisation

3 Demolition

• Dismantling
• Recycling
• Waste disposal

T_n

Figure 3.3 An environmental life cycle costing scheme

NO_x, dust and solid wastes. It is also recommended that a virtual rate is used when calculating the current value of future environmental costs, because future technology is assumed to be more environmentally effective than the current technology.

The coefficients $N(t)$ and N_s can either be calculated from known equations or taken from general financial calculation tables. The life cycle costs can be used both for optimisation and as a selection criteria between design alternatives and products.

The components of the ecological costs are: consumption of non-renewable raw materials and energy, the production of pollutants into air, soil and water, including CO_2, CO, SO_2, NO_x, dust, solid wastes and noise, and the loss of biodiversity.

Integrated environmental implications such as the 'greenhouse effect', 'acidity effect' or 'oxidants effect' can be used instead of individual environmental impact (Udo de Haes *et al.* 1999). It is also recommended that a virtual rate is used when calculating the current value of future environmental costs, because future technology is assumed to be more environmentally effective than the current technology.

Proposed rates in the discounting of future costs into present values are as follows:

	Real interest rate to be used in	
	Financial costs (%)	*Environmental costs (%)*
Cost to society	2	1–2 (5)
Cost to firms and individuals	3–5 (10)	1–2 (5)

The rates in parentheses are extreme values, which should only be used in special cases. For the enviromental calculations the rate describes the estimated speed of development in environmental efficiency of technologies. An ordinary rate of development would be represented by 1 per cent, 2 per cent in the case of rapid development being estimated and 3 per cent for special cases where rapid changes in technology are assumed. As examples, the proposed rates lead to the following values of the coefficient $N(t)$:

Time from construction (years)	$N(t)$, the annual rate r being			
	1%	*3%*	*5%*	*10%*
10	0.90	0.74	0.61	0.38
20	0.82	0.55	0.38	0.15
30	0.74	0.41	0.23	0.057
50	0.61	0.23	0.087	0.009
100	0.37	0.052	0.009	0.000

There are also aspects, such as biodiversity, which cannot be calculated numerically. These can only be evaluated qualitatively and descriptively. In qualitative evaluation the following methods can be applied:

* classified costs and values: division in cost and value classes
* estimated qualitative costs and values: estimation by calculation or experience
* normative costs and values: limit values presented in norms and standards

- ranking of the importance of costs and values: ranking list based on the weighting of different costs and values in numerical calculations
- intuitive valuation: valuation based on intuitive valuation of one or more people involved in the design.

The financial life cycle cost components of structural assemblies should be presented in future, for example in the suppliers' product descriptions. They can also be collected into a contractor's statistics database or into general design manuals, such as *Life Cycle Costing for Design Professionals* (Kirk and Dell'Isola 1995).

Environmental costs are calculated in the detailed design phase with the same equations (3.1)–(3.3) as in the conceptual design phase, but with more precision, using suppliers' product information, contractors' environmental statistics or general environmental statistics, such as manuals and computer tools (Lippiatt 1998a, Lippiatt 1998b).

3.2.3 Design life

Design life is a specified time period, which is used in calculations. Ordinary design life is 50 years (CEN 1991) for buildings and 100 years for civil engineering structures. In special cases even longer design life cycles can be used. However, after 50 years the effect of increased design life cycle is quite small and it can be estimated as the residual value at the end of the calculation life cycle. Temporary structures are designed for a shorter design life, which will be specified in each individual case. In ENV 1991–1 the following classification of design life is presented:

Class 1: 1–5 years	Special case temporary buildings
Class 2: 25 years	Temporary buildings, e.g. stores buildings, accommodation barracks
Class 3: 50 years	Ordinary buildings
Class 4: 100 years	Special buildings, bridges and other infrastructure buildings or where more accurate calculations are needed, e.g. for safety reasons
Class 5: over 100 years	Special buildings e.g. monuments, very important infrastructure buildings

3.2.4 Estimation of recycling efficiency

The components of the environmental profile of basic materials already include recycling efficiency – the environmental costs of materials recycling. It is important to realise that in design, the recycling potential of building components, assemblies and even technical systems must be considered. The higher the recycling is in the building's heirarchy, the higher the ecological and economic efficiency of the recycling will be.

At the conceptual phase, the estimation of recycling efficiency is typically using the recycling efficiency indices of the structural assemblies or components. The estimation of the recycling efficiency indices of different alternative solutions of structural systems or assemblies can be carried out by applying the scheme presented in Table 3.4.

Rough estimates of the recycling efficiency are:

1 Recycling of materials:
   - consumption of raw materials 0.5–0.9
   - consumption of energy 0.0–0.5
   - production of pollutants 0.1–0.7

2 Recycling of structural or installation assemblies or components:
   - consumption of raw materials 0.7–0.9
   - consumption of energy 0.7–0.9
   - production of pollutants 0.8–0.9

3.2.5 Estimation of residual value

The residual value of different modules of the building at the end of design life time (e.g. 50 years) can be estimated with the same statistics and methods that are used in taxation and insurance contexts. The producers' service life can also be used in calculating the residual value with the equation

residual value = (value at time t = 0) × ((characteristic service life / design life time) – 1)   (3.1)

3.2.6 Energy costs of the building

Basically the calculation of life cycle energy costs is part of the calculation of environmental life cycle costs. However, because of the great importance (typically about 80 per cent of all environmental burdens) of energy in life cycle environmental costs, and because of its considerable impact also in financial life cycle costs of the building, the methods in calculations of life cycle energy are outlined briefly below.

Energy costs are determined at the draft design phase depending on the functional and technical specifications of the building. The following classification of buildings can be roughly used for definition of, and decisions about, the energy costs of the building at sketch and draft design phase:

Class 1. Standard level. Heating/cooling energy costs appropriate to current standards of the country or region. Heating/cooling energy consumption is typically 120–180 kWh/m^2y and total energy consumption is 150–200 kWh/m^2y. The area being calculated as a living floor area.

Table 3.4 Estimation of the recycling efficiency coefficient k_r

Recycling factor	Basic value and residual changes of the recycling factor at the level of			
	Building	*Asssembly*	*Component*	*Material*
Basic value	0.6	0.5	0.4	0.2
Separability				
– High	+0.2	+0.2	+0.2	+0.2
– Moderate	0.0	0.0	0.0	0.0
– Low	–0.2	–0.2	–0.2	–0.1
Moveability				
– High	+0.05	+0.05	+0.05	+0.05
– Moderate	0.00	0.00	0.00	0.00
– Low	–0.05	–0.05	–0.05	–0.050
Reintegrability				
– High	+0.1	+0.1	+0.1	+0.1
– Moderate	0.0	0.0	0.0	0.0
– Low	–0.1	–0.1	–0.1	–0.1
Final value	0.25 ... 0.90	0.15 ... 0.85	0.10 ... 0.75	0.00 ... 0.55

Class 2. Reduced energy level. Heating/cooling energy consumption of 60–80 kWh/m²y and total energy consumption (including lighting and other building services systems use) less than 100 kWh/m²y.

Class 3. Low energy level. Heating/cooling energy consumption of 30–40 kWh/m²y, and the consumption of total energy (including lighting and building services systems use) less that 60 kWh/m²y.

Class 4. Zero energy level. Heating/cooling energy consumption is zero. This needs active solar energy gain or some other natural energy source such as from the earth.

Class 5. Energy gain building The gain from solar or other natural energy is more than that needed for heating/cooling and building services systems.

3.3 Generic systematics from requirements into technical specifications

3.3.1 Classification of users' requirements

The following is a systematised classification of the functional requirements (Leinonen and Huovila 2000), applying the systematics of CIB Master List. This list can be used as a check-list in requirements analysis. This functional requirements list is as follows:

A Performance
- A1 Conformity
  - A1.1 Core processes
  - A1.2 Supporting processes
  - A1.3 Corporate image
  - A1.4 Accessibility
- A2 Location
  - A2.1 Site characteristics
  - A2.2 Transportation
  - A2.3 Services
  - A2.4 Loadings to immediate surroundings
- A3 Indoor conditions
  - A3.1 Indoor climate
  - A3.2 Acoustics
  - A3.3 Illumination
- A4 Service life and deterioration risks
  - A4.1 Service life
  - A4.2 Deterioration risks
- A5 Adaptability
  - A5.1 Adaptability in design and use
  - A5.2 Space systems and pathways
- A6 Safety
  - A6.1 Structural safety
  - A6.2 Fire safety
  - A6.3 Safety in use
  - A6.4 Intrusion safety
  - A6.5 Natural catastrophes
- A7 Comfort

B Cost and environmental properties
- B1 Life cycle costs
  - B1.1 Investment costs
  - B1.2 Service costs
  - B1.3 Maintenance costs
  - B1.4 Disposal and value
- B2 Land use
- B3 Environmental burdens during operation
  - B3.1 Consumption and loads, building
  - B3.2 Consumption and loads, users
- B4 Embodied environmental impacts
  - B4.1 Non-renewable natural materials
  - B4.2 Total energy
  - B4.3 Greenhouse gases
  - B4.4 Photochemical oxidants
  - B4.5 Other production related environmental loads
  - B4.6 Recycling

C Requirements of the process
    C1 Design and construction process
        C1.1 Design process
        C1.2 Site operations
    C2 Operations
        C2.1 Usability
        C2.2 Maintainability

3.3.2 Phases of performance-based design procedure

Integrated lifetime engineering methodology aims to regulate optimisation and ensure life cycle human conditions, economy, cultural compatibility and ecology through technical performance parameters, as presented in Figure 0.1. With the aid of lifetime engineering we can thus control and optimise the human conditions (safety, health and comfort), the financial costs and environmental costs. Beside these, social aspects also have to be taken into consideration.

The phases of performance-based planning and design are:

1 analysis of the functional and performance requirements of the user
2 analysis and optimisation of the performance properties of the structure based on the functional requirements of the user
3 specification of the technical properties of the structure, based on the performance properties.

The procedural schema is as presented in Figure 3.4.

3.3.3 Quality function deployment method

The quality function deployment (QFD) method is related to linear programming methods which were widely used in the 1960s in industrial product development. As currently formulated QFD was developed in Japan and was first used in 1972 by the Kobe Shipyard of Mitsubishi Heavy Industries (Zairi and Youssef, 1995).

After that QFD has been increasingly used in Japan and since the 1980s also in the USA, Europe and worldwide. Until now its main use has been in the mechanical and electronics industries, but applications also exist in construction.

QFD is used to translate requirements into either functional or technical specifications. Thus QFD can serve as an optimising or selective linking tool between requirements and specifications. It can be used both for product development or for the design of individual buildings. Fundamental objectives of QFD are:

- identification of customers' needs
- interpreting customer needs first into the functional and then into the technical specifications of the building
- optimising the technical solutions with regard to requirements
- selection between different design alternative

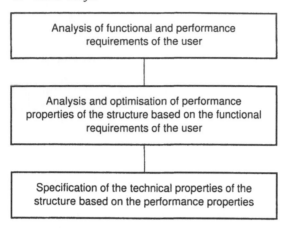

Figure 3.4 Procedural schema from functional
requirements to technical specifications

Put simply, the QFD method requires the building of a matrix between requirements (the 'whats') and design specifications (the 'hows'). The priorities of requirements, priorities of specifications, and the correlation between requirements and correlation between specification can also be undertaken. In practical design applications it is limited to a few key requirements and specifications in order to maintain good control of variables and to avoid wasting efforts on secondary factors. At product development a more detailed application can be used. The principle of key attributes and parameters can be applied differently to each type of design case in order to get an optimal use of QFD method. A model table – a 'house of quality' – is presented in Figure 3.5.

In industrial engineering, manufacturing companies have successfully applied concurrent engineering tools, such as quality function deployment (Akao 1990) to include customers' needs for product features into design at early stages of development, to integrate concurrent design of products and related processes, and to consider all elements of the product life cycle. Customer-oriented 'champion products' may also be priced higher than their competitors, and still become market leaders. In spite of its success stories in other industries during the past decade, QFD has rarely been applied in construction. However, examples from Japan, the United States, Finland, Sweden and Chile show it also has potential in building design. (Nieminen and Huovila 2000).

QFD provides an empty matrix ('house of quality', Figure 3.5) to be filled with customer requirements and their importance in the rows along the left-hand side, and properties of the solutions in the columns along the top portion. The centre describes the matrix-relationship of requirements and corresponding solutions. The importance measures (weighting factors) are at the bottom, and the right- hand side of the box shows the evaluation of competing alternatives.

The following phases of a construction process are identified as potential for QFD implementation in construction, Figures 3.5 and 3.6 a and b:

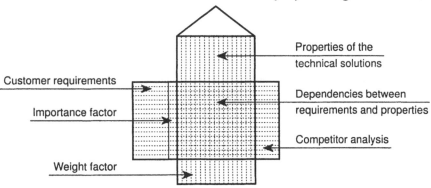

Figure 3.5 A 'house of quality' (Nieminen and Huovila 2000)

1 programming: customers' requirements for building and design objectives
2 design: design objectives and construction drawings
3 production: construction drawings and production plans
4 construction: production plans and construction phases.

Priorities of design specifications

The following procedure can be applied when using QFD to analyse functional requirements against owners' and users' needs, technical specifications against functional requirements, and design alternatives or products against technical specifications:

1 identify and list factors for 'what' and 'how'
2 evaluate and list priorities or weighting factors of 'whats'
3 evaluate correlation between 'whats' and 'hows'
4 calculate the factor: correlation times weight for each 'how'
5 normalise the factor: 'correlation times weight' of each 'how for use as a priority factor or weighting factor of each 'how' at the next stage.

A schedule for analysis between primary functional attributes and technical specifications can be presented as shown in Figure 3.6b.

Technical specifications priorities can again be used as weighting factors for comparisons between different design alternatives and products regarding owners' and users' demands and requirements. The following two case studies illustrate similar schemes (Nieminen and Huovila 2000).

3.3.4 Examples

CASE STUDY 1: VILLA 2000

The design of Villa 2000 was based on teamwork, where each designer could influence decision-making in other design fields outside his own design

	Correlation between 'hows'					
	Primary functional attributes					
	1	2	3	4	5	
Primary demands	Correlation between 'hows' and 'whats' c					Priorities of demands, *p*
1. Financial costs	1.0	0.5	0.2	0.5	0.3	9
2. Functionality	0.0	1.0	0.0	0.1	0.0	10
3. Environmental costs	0.0	0.0	1.0	0.0	0.0	8
4. Health	0.0	0.1	0.0	1.0	0.2	10
5. Aesthetics	0.0	0.0	0.0	0.2	1	8
Σ (*c* × *p*)	9.0	15.6	9.6	15.7	12.7	Σ = 62.6

Priorities of functional attributes

$$= c \times p/ \Sigma(c \times p)$$

0.14	0.25	0.15	0.25	0.20

Primary functional attributes
1 Life cycle financial costs   4 Life cycle environmental costs
2 Life cycle functionality   5 Health, aesthetics and comfort
3 Life cycle maintenance

Figure 3.6a Demands and their priorities

	Correlation between 'hows'									
	Primary technical specifications									
	1	2	3	4	5	6	7	8	9	10
Primary functional attributes	Correlation between 'hows' and 'whats', c									Priorities of functional attributes
1	0.3	0.2	0.2	0.2	0.0	0.0	0.0	0.0	0.0	0.3	9
2	0.5	0.3	0.3	0.3	0.0	0.2	0.2	0.1	0.3	0.3	10
3	0.5	0.5	0.4	0.1	0.0	0.1	0.1	0.1	0.2	0.8	9
4	0.2	0.1	0.1	1.8	1.0	0.0	0.0	0.0	0.1	0.3	8
5	0.0	0.0	0.0	0.4	0.2	0.3	0.3	0.1	1.0	0.5	10
	13.8	10.1	9.2	16.1	10.0	18.0	5.9	2.9	15.6	20.3	
Σ (*c* × *p*)	Priorities of functional attributes = *c* × *p*/ Σ (*c* × *p*)										Σ = 121.9
	0.11	0.08	0.08	0.08	0.13	0.15	0.005	0.02	0.13	0.17	

Primary functional attributes

1	Product structuring for changes and reuse	6	Hygrothermal specifications
2	Service life specifications	7	Acoustic specifications
3	Static and dynamic specifications	8	Fire safety specifications
4	Energy specifications	9	Health specifications
5	Environmental specifications	10	Maintainability and repairability

Figure 3.6b Functional attributes and their priorities

responsibilities. A design briefing process was organised to acertain the owners' requirements for the building. The QFD method was used experimentally to set design guidelines for Villa 2000. The IEA Task 23 criteria were expressed in a form of performance requirements and they are given weights (on a scale of 1 to 5) depending on their importance (IEA 1998). The potential design solutions are then created in the form of properties and their correlation with the requirement is given (on a scale of 0, 1, 3 or 9). The QFD spreadsheet summarises numeric values of the properties at the bottom of the matrix by multiplying the correlation with their weights, so that high values indicate high priorities. The user may then select the most important properties as a basis for the next phase of development.

The exercise was conducted with a group of ten experts from different backgrounds. The following objectives were set for the working session:

- to share common understanding of the performance-based objectives of the end product (a building to be designed and constructed)
- to prioritise the project objectives
- to strive for innovative design solutions that meet these objectives.

The first matrix (Figure 3.7a) shows the main selected objectives of a housing project (adaptability, indoor conditions, economy, environment friendliness, constructability and architecture) used as a basis for building design. The second matrix (Figure 3.7b) shows the structured approach in the design process based on the selection made in phase 1. The importance of the whole design and construction process was recognised as the key to fulfil the requirements, and the functionality and adaptability of the house as the key to future housing.

CASE STUDY 2: NURSERY SCHOOL

The second QFD example was to set the project objectives with a view to the building users' needs and requirements and to show how the chosen criteria and the users' view affect the results. The QFD matrix was used to acertain, record and verify the clients' requirements and to test the dependency between the requirements and the properties of the proposed building concept.

The project used in the test was a nursery school for about 100 children built in the year 2000. The design process of the building was finished towards the end of 1999, based on an architectural competition. The Merituuli nursery school was built in a new suburban housing area, formerly an industrial area, where the basic infrastructure was already developed (streets, access to main roads, district heating network, etc.). The location of the area is very close to the city of Helsinki with good public access to the city, which has made the area very popular, especially among young families. This has also grown to be a design feature for the nursery school building and its connection to the surrounding housing area.

The building serves as a nursery school during the daytime, and in the evening as a meeting point for community activities. The total building area is 1260 m² in a single storey. The owner of the building is the City of Helsinki, and the

PHASE 1 Properties \ Requirements	adaptability	resale value	indoor conditions	attractiveness	economy	autonomy	friendliness to environment	futurity	respond to environment	good indoor climate	habitability	constructability	identity	total ecology	architecture	simple user interfaces	recyclable fair house	transferability	dismountability		Importance factor
functionality Utilisability	9	9	9	9	3	9	3	0	9	0	9	8	1	1	8	9	3	1	0		5
Adaptability	9	3	8	9	3	1	9	3	9	0	0	1	1	9	8	1	9	9	9		2
Maintainability	3	3	3	3	9	9	9	0	9	0	3	8	0	9	1	3	1	1	1		2
environmental loading Operation	9	3	9	3	9	9	9	1	1	9	9	8	0	9	8	0	0	0	0		4
Construction	0	8	8	3	3	0	9	0	0	0	0	9	1	9	1	0	9	9	9		2
resource use Energy	9	3	9	3	9	9	9	9	0	9	9	8	3	9	8	0	1	1	1		5
Water	9	1	8	1	3	9	9	3	1	0	0	8	0	3	8	1	0	0	0		1
Materials	3	9	9	3	9	1	9	9	9	0	9	9	9	9	3	0	9	9	9		1
life cycle cost Investment cost	9	9	3	3	9	3	8	0	0	3	3	9	1	0	8	1	3	3	3		3
Operating cost	9	9	1	3	9	9	9	3	0	3	1	8	3	3	9	9	3	3	3		4
Maintenance cost	9	9	3	9	9	9	9	0	9	3	8	3	3	9	3	3	3	3			2
indoor quality Acoustic comfort	9	9	9	9	8	0	8	9	9	0	0	3	3	0	9	0	0	0	0		2
Thermal comfort	9	9	9	9	8	0	3	9	9	9	3	3	0	9	3	0	0	0			3
Lighting	9	9	9	9	3	9	3	9	9	9	9	8	1	9	1	0	0	0			4
Indoor climate	3	9	9	9	8	0	3	9	9	9	9	9	1	8	0	0	0	0			5
architecture Architecture	9	9	9	9	9	3	8	9	9	3	0	9	0	9	1	3	3	3			3
Weight factor	393	355	322	307	205	273	250	250	248	246	241	182	180	179	169	118	112	102	97	0	4317
Weight factor %	9%	8%	7%	7%	7%	6%	6%	6%	6%	6%	6%	4%	4%	4%	4%	3%	3%	2%	2%	0%	100%
Votes	4	1	3		2	1	3				1		2		4	1	1				
Selected	x		x		x		x					x			x						

Figure 3.7a　Design objectives for a housing project, phase 1 (Nieminen and Huovila 2000)

PHASE 2 Properties \ Requirements	SPACE	PROCESS	STRUCTURES	MATERIALS	ENERGY	EQUIPMENT	Importance factor (P1)
adaptability, simple interfaces, re-usable fair house	9	9	9	3	3	1	3
indoor conditions, responds to the environment	9	9	9	9	9	9	4
economy, resale value	9	9	9	9	9	9	1
environmental, autonomy, total ecology	9	3	9	9	9	9	5
constructability	1	9	3	1	1	1	3
architecture	9	9	3	9	1	0	2
Weight factor (P1)	138	134	133	120	104	95	724
Weight factor %	19 %	19 %	18 %	17 %	14 %	13 %	100 %

Figure 3.7b　Design objectives for a housing project, phase 2

Construction Management Division (HKR) of the City of Helsinki constructed the building.

In a number of development sessions arranged between the client and VTT (Technical Research Centre of Finland) at the beginning of the project and, later between the designers, project management and VTT, the project goals and limits were discussed and the requirements were set. The decision-making in the project was tested against the main criteria adopted from the IEA Task 23 framework.

The results of the design briefing sessions were used as building owner-defined sub-requirements in compiling the QFD matrix (Figure 3.8).

According to the QFD results, the main properties of the nursery school building corresponding to the given requirements are district heating, bicycle access to the site, cleanable ventilation ductwork, multi-use playrooms for children and low-energy building envelope.

Environmental goals of the project were prioritised as the most important properties. Even though the builder has an environmental programme to support sustainable construction, it is not surprising that the requirements dealing with functionality or air quality in a nursery school dominated the pre-design process.

QFD method was used to evaluate the proposed technical design solution and to compare the design with typical existing nursery schools. A low-energy concept for the nursery school was also developed and analysed accordingly. The proposed solution (basic design, Figure 3.9) shows improvements compared to typical nursery schools. By improving the energy efficiency of the building, both user-friendliness (functionality, indoor climate) and environmental properties and life cycle costs were improved.

3.4 Multiple-attribute analysis, optimisation and decision-making (MADA)

3.4.1 Background to the method

The main phases of the life cycle design procedure are: analysis of the actual requirements, interpretation of the requirements into technical performance specifications of structures, creation of alternative structural solutions, life cycle analysis and preliminary optimisation of the alternatives, selection of the optimal solution between the alternatives and finally the detailed design of the selected structural system.

Multiple criteria optimisation is a central tool of structural designers in supporting decisions at the conceptual phase of design. The decisions, even in integrated design procedure, will be mainly based on empirical knowledge of the decision-makers, but supported by mathematical multiple-criteria analysis and multiple-variable optimisations (Baumann 1998). In this section the multi-attribute decision analysis method (MADA), which includes several variables is examined. (See also descriptions in Roozenburg and Eekels 1990, Lippiatt 1998a and 1998b, ASTM 1995, Norris and Marshall 1995.) The analysis made by Norris and Marshall (1995) shows, that in some cases where qualitative and quantitative financial and non-financial attributes are included in multi-attribute decision-making, a modified MADA method, non-traditional capital investment criteria (NCIC) is more suitable. NCIC developed by Boucher and Mac Stravic as a multi-attribute evaluation within a present worth framework and its relation to the analytic hierarchy process. This is also the case in structural design when dealing with comparison, selection and optimisation between several alternatives of structural systems, modules, components or materials. For this reason, only the NCIC method is presented here where it is applied to structural design.

PHASE 1 Properties / Requirements	district heat	bicycle access to site	cleanable ducts	multi-use playrooms	low energy envelope	mechanical ventilation + HR	changeable duct components	separated service space	super windows	floor heating	solar control	yard facing South	stimulating spaces, child scale	L-form	separated public evening use	ordinary windows	traditional envelope	radiators	Importance/Weight factor
LCC low investment cost	9	1	0	9	0	0	0	0	0	0	0	0	3	0	1	3	0		5
low service cost	9	1	9	3	9	0	0	3	9	3	0	0	3	0	0	0	0		4
low maintenance cost	9	1	9	3	0	0	0	9	0	0	0	0	1	0	0	0	3		1
resource use low electricity consumption	9	0	3	3	1	0	0	0	0	0	3	9	0	0	1	0	0		4
low water consumption	0	0	0	0	0	9	0	0	9	9	0	9	0	0	0	0	0		4
long service life	0	0	3	0	0	3	3	3	0	0	0	0	0	0	3	0	1		3
environmental loading low CO2, NOx, SO2 emissions	9	0	0	0	9	9	0	0	9	0	9	0	0	0	0	0			5
particles	0	9	9	0	3	9	9	0	9	0	0	0	3	0	0	0			5
existing infrastructure	9	3	0	0	0	0	0	0	0	0	0	0	3	0	0	0			1
archit quality home-like	0	9	0	9	0	0	0	1	0	3	0	3	1	0	0	1			3
attractive to children	0	3	0	9	0	0	0	0	9	9	0	3	9	1	0	0	0		4
public-service building	3	9	3	0	3	3	3	1	1	0	0	0	1	9	0	0	0		1
indoor quality air purity + emissions	0	9	9	0	0	3	3	0	0	0	0	0	0	0	0	0			5
high thermal quality	0	0	9	0	9	9	0	0	9	9	0	0	0	0	9	0	1		3
illumination	0	9	9	0	9	9	0	0	9	0	0	3	9	0	3	0	1		5
echoing	0	0	0	0	0	0	0	0	0	0	0	0	0	0	0	0			1
low HVAC noise	0	0	1	0	3	0	0	9	0	1	0	0	0	0	0	0			2
functionality user access to site	0	9	0	0	0	0	0	0	0	0	0	0	3	0	0	0			4
service access	0		0	0	0	0	0	9	0	0	0	0	0	0	0	0			3
safety in use	3	3	0	0	0	0	3	9	0	3	0	3	3	0	0	0	0		5
evening use	0	9	0	0	0	0	0	0	0	0	0	0	9	0	0	0			1
high adaptability	0	0	1	9	1	0	9	3	1	3	1	0	1	9	0	0	0		
Weight factor (P f)	207	189	180	153	136	135	132	130	111	106	107	99	99	69	42	19	12	0	1913
Weight %	9%	9%	7%	7%	7%	7%	6%	6%	5%	5%	5%	6%	2%	1%	1%	1%	0%	0%	100%
Selected	x	x	x	x		x	x	x		x									

Figure 3.8 Design priorities for a nursery school (Nieminen and Huovila 2000)

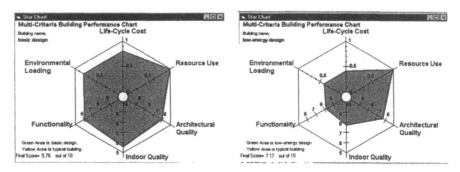

Figure 3.9 Analysis of two design solutions and comparison to properties of a typical nursery school building

3.4.2 Multi-attribute decision analysis (MADA) method and its modification NCIC

In the integrated life cycle design of structures the structural systems, modules, components and materials have to be analysed for selection between alternatives. The NCIC method incorporates a paired comparison procedure together with a pre-specified approach for converting the comparisons into attribute weights. The method allows hierarchical description of problems in order to keep the number of pairwise comparisons manageable. In structural design the hierarchical levels are: structural system, modules, components and materials. The method can deal with numerical quantitative values of attributes and with descriptive qualitative values of attributes. One of the attributes is always measured in monetary units attributed to each performance gain, and these values are summed to yield the overall implied value of each alternative. These implied values can be used to select an alternative, to rank alternatives, or to screen alternatives. The NCIC method is compensatory, which means that a high performance relative to one attribute can at least partially compensate for low performance relative to another attribute. However, minimum performance requirements have to be met.

3.4.3 MADA procedure in structural design

The MADA analysis can be used as a tool for multi attribute or multi-criteria optimisation and decision-making MCDM. MCDM includes analysis, comparison and selection procedures during the following phases:

1 definition of hierarchical levels to be dealt with
2 definition of attributes and their weights
3 definition of structural alternatives to be compared on each level of hierarchy
4 calculation of numerical score and a total value of each alternative
5 calculation of the incremental value of each alternative from each attribute
6 calculation of total value of each alternative
7 coupled comparisons of alternatives with the aid of their numerical core attributes
8 ranking of the alternatives primarily in regard to numerical core attributes but also rin regard to incremental value attributes or total value
9 selection of the best alternative for further design.

It is obvious that all minimum requirements of general laws, norms and standards must be fulfilled.

3.4.3.1 Attributes and their weights

Typical attributes and sub-attributes of structures are presented in Table 3.5. In special cases the attributes can be modified and changed if these general attributes do not reflect the actual needs. Life cycle monetary costs and functionality as primary attributes are always numerical. Therefore functionality must be evaluated numerically using the QFD method.

Because of two existing primary attributes (monetary life cycle cost and functionality) we can apply their ratio: functional efficiency (functionality/cost) as a primary attribute in comparisons. In that case pure values of life cycle cost and life cycle functionality can be treated as incremental values.

Table 3.5 General attributes and sub-attributes of structures

Attribute	Subattribute
1. **Life cycle monetary costs**	**Construction cost** Energy cost during design service life Maintenance cost during design service life Repair costs during design service life Changing costs during design service life Renewal costs during design service life Recycling cost Disposal cost
2. **Life cycle functionality**	**Functionality for the first user** **Flexibility for changes of building services** Flexibility for spaces Flexibility for changes in performance of structures
3. Life cycle maintenance	**Reliability in operation in normal and abnormal conditions** Ease, frequency and staff requirements of maintenance
4. Life cycle environmental costs	**Consumption of energy:** renewable/ non-renewable/fossil Consumption of raw materials: renewable/ non-renewable Production of pollutants and disposals into air, soil and water
5. Health and comfort	**Internal air quality (emissions)** Working conditions during construction Hygrothermal quality of internal conditions Acoustic and visual privacy and convenience Visual quality and aesthetics

Note:

Commonly dominant attributes or subattributes that dictate about 70–80 per cent of actual life cycle costs or value of buildings are shown in **bold**.

The life cycle monetary and environmental costs are mainly calculated numerically with relevant calculation methods. Functionality as a primary attribute must be evaluated numerically using the QFD method. The other attribute values are mainly presented descriptively, but also partly numerically, and the values are divided for example into four classes: very high, high, moderate and low. Some of the values, like environmental costs, are negative. Each primary attribute is evaluated as a weighted sum of its sub-attributes for calculating the increment values of each attribute in the case of each alternative.

3.4.3.2 Hierarchical levels

Life cycle optimisation increases the need for a hierarchical system approach, where buildings or other structural systems are considered as different entities,

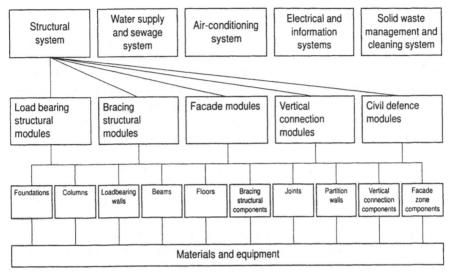

Figure 3.10 Hierarchy of the structural system of buildings (Sarja 1989)

starting from sub-buildings, technical systems or modules and ending at components and their details (Sarja 1989, Sarja and Hannus 1995).

The comparisons and selections are generally made first on a complete building level, modularising the building into technical systems.

In structural design the following hierarchical levels are used:

1 structural system
2 structural module: such as the loadbearing frame (or superstructure), foundations, envelope, partition walls, connection modules (staircases), roofing
3 structural components: such as a beam, a column, a slab etc.
4 materials and details finishing, plastering, water proofing membranes of floors and roofs, joints, floorings, paintings, furniture etc.

The hierarchy of the structural system is presented as an example in Figure 3.10.

3.4.3.3 Alternatives at each level of the hierarchy

The first phase for the definition of structural alternatives is the creative design work phase and will produce ideas for the structural system and its key modules. In building design the key modules can often be used as a start for the entire structural system. The key modules are usually the loadbearing frame, including the floors, and the envelope. At the creative phase the attributes and their weights already serve as an intuitive background. Systematic methods of creative product development can also be applied. The result of this definition are a set of probably between two and four alternative structural systems and their key modules.

3.4.3.4 *Numerical score and total implied value of each alternative*

The first step in the selection procedure is the calculation of numerical score v_{i1}, where i is the number of the alternative. The numerical score v_{i1} is always calculated in present value monetary units over the entire design service life. Thus it represents the first attribute: 'financial cost'. This means that financial cost is isolated as a dictating factor, and the other factors are treated as incremental values. The cardinal numerical score v_{i1} is calculated as a present value of life cycle monetary costs using separately the methods described here. The present value can be calculated either directly by the current value method, or through annual values converting them later into present value over the design service life period.

Because there are two existing primary attributes (financial life cycle cost and functionality) we can apply their ratio: functional efficiency (functionality/cost) as a primary attribute in comparisons. In this case pure values of life cycle cost and life cycle functionality can be treated as incremental values. The life cycle monetary and environmental costs are mainly calculated numerically with relevant calculation methods. Functionality as a primary attribute can be evaluated numerically using the QFD method, or it can be evaluated.

Each attribute j other than the financial cost attribute, contributes an independent increment of value v_{ij}, to the total value of the alternative i. The increment of value v_{ij} can be either positive or negative. The costs related to each attribute are calculated as negative values and the benefits as positive values. Usually only the monetary value v_{i1} and some other attributes (often only some sub-attributes), like environmental costs, are calculated numerically. The other attributes are expressed as descriptive qualitative values: 'very high', 'high', 'moderate' and 'low'. The total implied value V_i of each alternative i is calculated as the sum of the value increments associated with its performance relative to each of the n leaf attributes, applying the linear equation

$$V_i = \sum_{j=1}^{n} v_{ij} \tag{3.4}$$

V_i is used to identify a subset of most preferred alternatives, to rank the alternatives, and to select the single most preferred alternative. The weight coefficients of the alternatives are defined separately for each alternative. Thus, the weight coefficients are not necessarily exactly constant across alternatives, because the importance of each attribute may be different between alternatives. Therefore the weight coefficients of this method do not characterise exactly the general relative importance of the attributes.

3.4.3.5 *Incremental values of attributes*

At the second step, the incremental values of each attribute j of each alternative i have to be normalised, because they have different qualitative or quantitative measures. This is done comparing the value of each attribute to the value of a chosen baseline alternative 1, which usually represents the most ordinary alternative. Thus, the modified (normalised and weighted) incremental value v_{ijm} is calculated with the equation

$$v_{ijm} = v_{ij} / v_{1j} \tag{3.5}$$

3.4.3.6 Total implied value of alternatives

The modified total implied value V_{im} of alternative i is calculated with the equation

$$V_{im} = \sum_{j=1}^{n} V_{ijm} \tag{3.6}$$

Each normalised and weighted incremental value v_{ijm} thus varies between $v_{ijm} = \pm (0, k_{ij})$, where k_{ij} is the weight coefficient of the attribute j of alternative i.

3.4.3.7 Pairwise comparisons of alternatives

At the first step, at each alternative the pairwise comparison of attributes is made ranking the importance of each attribute in that alternative. This procedure produces the weight coefficients between the attribute of each alternative.

3.4.3.8 Retrofitting alternatives

During the selection procedure some alternatives may show a need for changes in order to improve performance and thus the value. This retrofitting is defined as modifying the alternative in order to improve its performance. After retrofitting the primary numerical score value and the total implied value is calculated again.

3.4.3.9 Selection of the alternatives for further design

The selection can be made by including the comparison of monetary life cycle costs together with multiple, conflicting non-monetary attributes of alternatives. The NCIC method is designed to make explicit the implied monetary value of the non-financial benefits. The results of an NCIC analysis can also be integrated into traditional economic analysis. Often the factors of the environmental profile are quite well interrelated, which makes comparisons and optimisations easy (United States Environmental Protection Agency (USEPA) 1990, Norberg-Bohm, Clark, Bakshi, Berkenkamp, Bishko, Koehler, Marrs, Nielsen and Sagar, 1992). Sometimes different components have different optimum points. In such cases, a valuation between the components with different measurement systems (for example, costs are measured in currency, whereas air pollution is measured in tons of CO_2 and NO_x) must be done by using suitable weighted coefficients and normalised values for the components (Udo de Haes, Jolliet, Finnveden, Hauschild, Krewitt, and Müller-Wenk 1999, Sarja 2001). Normalisation can be done by comparing all alternatives with a commonly-known reference alternative and dividing the corresponding values of each valuation component.

Complicated analysis can, in design, be done for very large and important building projects and in the development process of entire building concepts or building products. In the design of ordinary buildings the method can be used in simplified ways. The numerical value of the cardinal numerical score (the present value of costs over design life time period) has always to be calculated using general characteristic values at a level of functional units like spaces and structural modules. A simple calculation of total implied value of each alternative on a hierarchical level corresponding to the monetary cost calculations can also be made. The selection can be based on partly intuitive and subjective decisions by the client or owner, although this should be supported by the analysis made by the structural designer. The alternatives selected subjectively may be:

- the alternative with lowest life cycle monetary cost, together with at least moderate total implied value and at least moderate values of important individual attributes
- the alternative with reasonable life cycle monetary cost together with highest total implied value
- the alternative with reasonable life cycle monetary cost together with high incremental values of most important attribute(s) and at least moderate incremental value of other attributes.

An example scheme of a decision table is presented in Table 3.6.

3.4.3.10 Values of variables

In analysis, optimisation and decision-making, the following types of values of variables are used:

1 calculated numerical financial and environmental costs and values
2 classified costs and values
3 estimated qualitative costs and values
4 normative costs and values
5 ranking of the importance of costs and values
6 intuitive valuation.

3.4.4 Integration of the design solutions

To integrate partial design solutions, which are produced by selecting between alternatives, applying the design tasks and methods described above, an additonal aggregation of performance parameters will be made followed by multi-attribute analysis and decision-making. The result will be a final definition of the technical performance parameters of structures and materials. The results will then be used as design criteria specifications in the subsequent finalising of the detailed design of the structure.

Table 3.6 An example decision-making table

	Primary attributes			Incremental values	
	Construction cost $E(0)$ life cycle cost (design life cost) $E_{tot}(t_d)$ Euro	Lifetime usability and changeability score / ranking (scale: 4–10)	Health, safety, convenience score / ranking (scale: 4–10)	Environmental cost relative ecoefficiency/ ranking	Cultural value score / ranking (scale: 4–10)
Limit or weighting factors specified by client	max $E(0)$ = 55 000 max $E_{tot}(t_d)$ = 105 000	Weighting factor = 10	Weighting factor =10	Weighting factor = 8	Weighting factor = 8
Alternatives Reference alternative 1	R = 52 000 E = 105 000	8.0 (3rd)	8.5 (3rd)	1.0 (3rd)	7.0 (3rd)
Alternative 2	R = 55 000 E = 103 000	9.0 (1st)	9.5 (1st)	1.3 (1st)	8.5 (2nd)
Alternative 3	R =53 000 E = 104 000	8.5 (2nd)	9 (2nd)	1.1 (2nd)	9 (1st)
Ranking[2]	All alternatives fulfil the requirements	1st Alternative 2 2nd Alternative 3 3rd Alternative 1	1st Alternative 2 2nd Alternative 3 3rd Alternative 1	1st Alternative 2 2nd Alternative 3 3rd Alternative 1	1st Alternative 3 2nd Alternative 2 3rd Alternative 1
Numerical score	–	80.0 90.0 85.0	85.0 95.0 90.0	56.0 62.0 52.5	56.0 68.0 72.0
Residual value after design life time (50 years) at present value	Alternative 1: 25 000 euro Alternative 2: 29 000 euro Alternative 3: 28 000 euro				
Final ranking and numerical score	Ranking 1: Alternative 2 = 315.0 Ranking 2: Alternative 3 = 299.5 Ranking 3: Alternative 1 = 276.5				

3.4.4.1 Aggregation of life cycle performance parameters

Because of the complexity of building systems, decisions between design alternatives for the building, as well as between its technical system, module and product alternatives, must be simplified so as to limit the number of parameters used in the final decisions. To achieve this a number of design parameters will be aggregated.

As described earlier, the final objective of the integrated life cycle design is optimised life cycle quality which consists of four dominant parameters (see Figure 1.2):

1 life cycle functionality
2 life cycle costs
3 life cycle ecology
4 life cycle human conditions

A practical procedure in design is to use these four quality parameters as primary criteria parameters when selecting between design alternatives and products. They are referrred to here as 'life cycle quality parameters', LCQ.

Each life cycle quality parameter consists of a set of technical performance parameters. The process of building the relation between a life cycle quality parameter and its technical performance parameters is called aggregating. The methods used in aggregating are presented in Table 3.7.

3.4.4.2 Aggregation procedure

An important phase of the optimisation or decision-making procedure is the aggregation of a large number of specific performance factors into the LCQ parameters presented in Table 3.7. The aggregation scheme of LCQ parameters is presented in Figure 3.11.

The weighting in aggregation of environemental parameters is made on the following levels:

1 global level
2 regional level
3 local level

Typical global factors, which always have a high weight are the consumption of energy and air pollution which include factors of global climatic change.

Typical regional factors are, for example, consumption of raw materials and water. In locations these factors are extremely important, but in others they, or some of them, have little relevance.

The weighting between safety, health and comfort can be made individually. Usually the weightings of health and safety are very high, while the weighting of comfort can vary more widely. In any case, health and safety must fulfil the minimum regulatory requirements, which usually are quite strong.

As an example we can take the weightings in northern Europe (Scandinavian countries). A weighting, where the factors of climatic change and air pollution ($CO_{2\,ekv}$, $SO_{2\,ekv}$ and ethane ekvivalent) are taken into account is widely used (Lindfors *et al.* 1995).

Table 3.7 Methods used in aggregating life cycle quality (LCQ) parameters from technical life cycle parameters

Life cycle quality parameter	Aggregation method	Criteria
1. Life cycle functionality (QFD)	Quality function deployment Normative minimum requirements and classifications	Functional efficiency
2. Life cycle financial costs (LCFC)	Life cycle costing	Economic efficiency (Normative minimum requirements and classifications)
3. Life cycle environmental costs (LCEC)	EPA Science Advisory Board (1990), Lippiatt (1998b) Harvard University Study (Norberg-Bohm *et al.* 1992)	Ecoefficiency Normative minimum requirements and classifications
4. Life cycle human conditions (LCHC)	Analysis of total volatile organic compounds (VOC) emissions (Radnünz 1998). Evaluation of fungal risk. Evaluation of risk of radioactive radiation from materials and from the earth. Evaluation of ventilating air quality. Evaluation of health risks to water quality.	Quality classifications of indoor air quality and other indoor air conditions. Quality classifications of acoustic performance. Normative minimum criteria and classifications of safety, health and comfort.
5. Overall Life Cycle Quality	Multi-attribute decision-making	Life cycle quality (LCQ)

3.4.4.3 *Normalised ecoefficiency parameter*

The general aggregated environmental life cycle environmental cost (LCEC) value, which is described above, can be used in calculating the normalised ecoefficiency parameter (ECOEFF). ECOEFF can be calculated as a ratio between LCEC of a reference object (product, design solution, building concept, production method etc.) and the LCEC of the actual object, using the equation (Sarja 2001).

$$ECOEFF = LCEC_{ref} / LCEC_{actual} \qquad (3.7)$$

where

ECOEFF   is the normalised ecological efficiency parameter

$LCEC_{ref}$ life cycle environmental cost parameter LCEC of the reference object

$LCEC_{actual}$ life cycle environmental cost parameter LCEC of the actual object

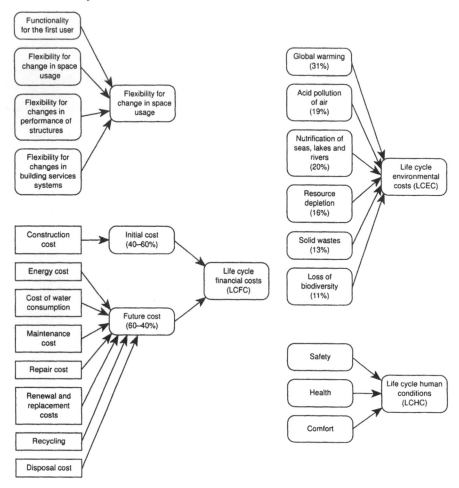

Figure 3.11 Aggregation scheme for LCQ parameters

3.5 Service life planning

3.5.1 Objective and principles

The objective of service life planning is to ensure that the service life of a building, its modules and components are functionally, technically, ecologically and economically optimised over the design life. However, the service life of different parts (modules or components) of a building are different. Also the decisive factor dictating service life varies, it may be defective performance, or functional, technical, ecological or economic obsolescence. From experience, we know that roughly 50 per cent of buildings are demolished because of defective performance and 50 per cent are demolished because of obsolescence. This dictating factor has to be identified separately at life cycle planning process for each part of the building, and for each alternative design or product. Once identified, the optimisation of the system through sequential comparisons between alternatives can take place. Because of the multiple categories of requirements, some feedback during the planning process is needed, which is why the process is partly iterative.

The service life of buildings and their subsystems, modules and components can be classified, for example by using the classification presented in Chapter 2. Another classification is presented in draft Standard ISO/DIS15686–1. The rough classification serves as a framework for service life planning and optimisation at briefing, conceptual and initial design phases, but more exact service life estimation is carried out at detailed design phase during selection between different products.

A model of modularised service life planning scheme is presented in Table 3.7. As can be seen, the target service life of some key modules like foundations and loadbearing frame can be defined as being longer than the design life of the building. This means that those modules retain a certain residual value after the design life, which can be taken into account in economic life cycle calculations both of financial and environmental costs.

Following the scheduled target service life, a multiple requirements life cycle optimisation can be made spearately for each building. The optimisation is carried out in order to determine optimal target service life values above the design service life of building, and optimal target service life inside the specified range of possible suitable values as presented in Table 3.8. A model applying the residual value method for modules having a characteristic service life above the calculated value for the building, and applying a method of multiplication coefficients for modules having several changes or renewals during the design service life of the building are presented in Table 3.9.

3.5.2 Service life dictated by defective performance

Defective performance is controlled in design with proper service life planning which results in realistic target service life for key modules of the building, and in detailed durability service life design which results in key structures which are owing an adequate performance over design service life.

3.5.3 Service life dictated by obsolescence

Obsolescence means the inability to satisfy changing functional, technical or economic requirements. Obsolescence can affect the entire building or just some of its modules or components.

Functional obsolescence is due to changes in functions and use of the building or its modules. This can even be when the building's location becomes unsuitable. This has commonly happened in country areas when people have moved into cities. In such cases the only possibility would be to dismantle and reassemble the building, which would require a demountable structural system. More common are changes in use which require changes in functional spaces or building services systems. This gives rise to a need for flexible structural systems, usually requiring

Table 3.8 A model of modularised service life planning scheme of a building with a design life of 50 years

Functional module	Ordinary target / design life due to				Target service life (CSL) (years) [1]
	Functional obsolescence (years)	Technical obsolescence (years)	Economic obsolescence (years)	Defective performance (years)	
Foundations	> 100	> 100	> 100	50– > 100	> = 50
Loadbearing frame	> 100 [2]	> 100	> 100	50– > 100	> = 50
Envelope / walls	> 50	> 50	50	30–200	30– > 50
Envelope / roof	50	50	50	15– > 100	20– > 50
Envelope / ground floor	> 50	> 50	> 50	30– > 100	> = 50
Envelope / windows	30–50	20–50	20–50	20–50	> = 30 [3]
Envelope / doors	> = 20	20–30	> 50	30– > 100	20– > 50 [3]
Partition floors [3]	1–50	20–50	> 100	> 100	> = 50 [5]
Partition walls (including doors) [3]	5–50	> 50	> 50	40– > 100	5–50 [3][4]
Bathrooms and kitchens [3]	20–30	15–30	25–40	20– >50	20–50 [5][2]
Building services systems	2–40	3–50	3–50	5–40	5–40 [5][3]

Notes

1　Target or design life will be defined inside the defined range through multiple requirements optimisation process.
2　Flexibility for changes of internal spaces is needed in periods of changes of other modules.
3　Recycleability of components is important.
4　Modules include compatibility between structures and installations.
5　Changeability of parts of module is highly needed.

long spans and minimum numbers of vertical loadbearing structures. Partition walls and building services systems which are easy to change are also required.

Changes in building services systems equipment are often the cause of technological obsolescence, but the structure can also be a cause when new products providing better performance become available. Typical examples are more efficient heating and ventilation systems and their control systems, new information and communication systems such as computer networks, better sound and impact insulation for floorings, and more accurate and efficient thermal insulation of windows or walls. Technological obsolescence can sometimes be avoided or reduced by estimating future technical development when selecting products. The effects of technical obsolescence can also be reduced through the proper design of structural and building service systems to allow easy change, renewal and recycling.

Economic obsolescence means that operation and maintenance costs are too high in comparison to new systems and products. This can partly be avoided in design by carefully minimising the operation and costs by selecting materials, structures and equipment which need minimum amounts of work and materials for maintenance and operation. Often this means simple and safe working products

Table 3.9 Residual value method and multiplication coefficient method in the optimisation of target service life of building modules (the building has design life of 50 years)

Functional module	Target service life (design service life, DSL)[1] (years)	Residual R_{50} after design service life of 50 years	Multiplication coefficient C during design service life of 50 years	Estimated time period for functional changes in actual building years	Multiplication coefficient C for further calculations
Foundations	≥ 50	(DSL–50) / DSL	1	–	1
Loadbearing frame	≥ 50[2]	(DSL–50) / DSL	1	–	1[2]
Envelope / walls	30– >50	0–(DSL–50) / DSL	1–2	–	1–2
Envelope / roof	20– >50	0–(DSL–50) / DSL	1–3	–	1–3
Envelope / ground floor	≥ 50	(DSL–50) / DSL	1	–	1
Envelope / windows	≥ 30[3]	0 to second residual value	2	–	2
Envelope / doors	20– >50[3]	0–(DSL–50) / DSL	1–2	–	1–2
Partition floors[3]	≥ 50[5]	(DSL–50) / DSL	1	–	1–2
Partition walls (including doors)[3]	5– >0[3 5]	0–(DSL–50) / DSL	1–10	CW = 5–50	1 <= CW <= 10
Bathrooms and kitchens[3]	20–50[5 3]	0 to second residual value	1– 3	CB = 20–40	1 <= CB <= 3
Building services systems	5–40[5 3]	0 to second residual value	1–10	CS = 5–50	1 <= CS <= 10[2]

1 Target or design life will be defined inside the defined range through the multiple requirements optimisation process.
2 Flexibility for changes of internal spaces is needed when other modules are changed.
3 Recycleability of components is important.
4 Modules include compatibility between structures and installations.
5 Changeability of parts of module is an important requirement.

which are not sensitive to defects and/or their effects. For example, monolith external walls are safer than walls. This selection is automatically included in several parts of integrated life cycle design and decision-making process.

3.6 Design for durability

3.6.1 Classification of durability design methods

Durability design methods can be classified starting from most traditional and ending in most advanced methods as follows:

1 design based on structural detailing
2 reference factor method
3 statistically calculated lifetime safety factor method
4 statistical durability design

3.6.2 Durability design with structural detailing

Structural detailing for durability is a dominant practical method which is applied to all types of materials and structures. The principle is to specify structural design and details as well as materials so that both deterioration effects on structures, and the effects of environmental impacts on structures, can be eliminated or diminished. The first of these is typically dominant when designing structures, such as wooden buildings, which are sensitive to environmental effects. The second principle is appropriate for structures which can be designed to resist even stronger environmental impacts, such as concrete, coated steel or wooden structures.

The methods and details for durability detailing are presented in current norms and standards, therefore this method is not described here.

3.6.3 Reference factor method

The reference factor method aims to estimate the service life of a particular component or assembly in specific conditions. It is based on a reference service life – in essence the expected service life in the conditions that generally apply to that type of component or assembly – and a series of modifying factors that relate to the specific conditions of the case (ISO/CD 15686–1). The method uses modifying factors for each of the following:

- A quality of components
- B design level
- C work execution level
- D indoor environment
- E outdoor environment
- F in-use conditions
- G maintenance level

Estimated service life of the component (ESLC).

$$ESLC = RSLC \times A \times B \times C \times D \times E \times F \times G$$

where RSLC is the reference service life of the component.

   The reference factor method is always an additive method, because reference service life always has to be known. The reference factor method is most often needed because the environmental exposure (environmental load onto structure) usually varies over a wide scale. Many parametric methods including the values of parameters in different conditions already exist: for example ISO 9223,1992 (F) contains methods and values for variables associated with steel structures
   For more details see ISO/CD 15686–1, where the method is described.

3.7 Limit state durability design

The statistical basis of this method follows the principles which are presented in (Sarja and Vesikari 1996). Although this method is presented and applied in detail for concrete structures, similar methods can also be applied to steel, wooden and masonry structures. Deterioration processes, dictating environmental loads, degradation factors and degradation calculation models are different for different materials.

3.7.1 Statistical service life

The simplest mathematical model for describing a 'failure' event comprises a load variable S and a response variable R. In principle the variables S and R can be any quantity and be expressed in any units, the only requirement is that they are commensurable. Thus, for example, S can be a weathering effect and R can be the capability of the surface to resist the weathering effect without too much visual damage or loss of the concrete reinforcement cover.
   If R and S are independent of time, the 'failure' event can be expressed as follows (Sarja and Vesikari 1996)

$$\{failure\} = \{R < S\} \tag{3.7}$$

The failure probability P_f is now defined as the probability of that 'failure':

$$P_f = P\{R < S\} \tag{3.8}$$

Either the resistance R or the load S or both can be time-dependent quantities. Thus the failure probability is also a time-dependent quantity. Considering $R(\tau)$ and $S(\tau)$ are instantaneous physical values of the resistance and the load at the moment τ the failure probability in a lifetime t could be defined as:

$$P_f(t) = P\{R(\tau) < S(\tau)\} \text{ for all } \tau \le t \tag{3.9a}$$

The determination of the function $P_f(t)$ according to Equation 3.9a is mathematically difficult. That is why R and S are considered to be stochastic quantities with time dependent or constant density distributions. By this means the failure probability can usually be defined as:

$$P_f(t) = P\{R(t) < S(t)\} \tag{3.9b}$$

According to the Equation 3.9b the failure probability increases continuously with time as schematically presented in Figure 3.12.

At a given moment of time the probability of failure can be determined as the sum of products of two probabilities: 1) the probability that $R < S$, at $S = s$, and 2) the probability that $S = s$, extended for the whole range of S:

$$P_f(t) = P_f = \int P\{R < S \mid S = s\} P\{S = s\} \tag{3.10a}$$

Considering continuous distributions the failure probability P_f at a certain moment of time can be determined using the convolution integral:

$$P_f = \int F_R(s) f_s(s) ds \tag{3.10b}$$

where $F_R(s)$ is   the distribution function of R,
        $f_s(s)$ the probability density function of S, and
        s the common quantity or measure of R and S.

Considering continuous distributions the failure probability P_f at a certain moment of time can be determined using the convolution integral:

$$P_f = \int F_R(s) f_s(s) ds \tag{3.10b}$$

where $F_R(s)$ is   the distribution function of R,
        $f_s(s)$ the probability density function of S, and
        s the common quantity or measure of R and S.

The integral can be solved by approximative numerical methods.

3.7.2 Lifetime safety factor method

In practice it is reasonable to apply the lifetime safety factor method when the formulation of the design procedure returns to the deterministic form. The lifetime safety factor method is analogous with the static limit state design. A comparison of static limit state design and durability limit state design is presented in Table

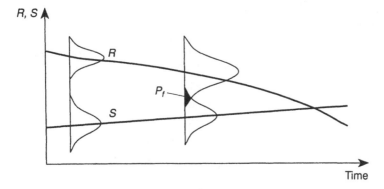

Figure 3.12 The increase of failure probability

3.10. It can be stated that durability design using the lifetime safety factor method is related to controlling the risk of falling below the target service, while static limit state design is related to controlling the reliability of the structure against failure under external mechanical loading.

The lifetime safety factor method is always combined with static or dynamic design and aims to control the service life, while static and dynamic design controls the loading capacity.

The lifetime safety factor can be calculated with the equation 3.23 (Sarja and Vesikari 1996)

$$t_d = \gamma_t t_g \qquad\qquad (3.12b)$$

where

    t_d is the design service life
    γ_t the central lifetime safety factor
    t_g the target service life.

Equations, which can be chosen as an alternative for calculating the design life t_d are (Sarja and Vesikari 1996, Sarja 2001):

$$t_d = t_0 / \gamma_{t0}$$

or $\qquad\qquad\qquad\qquad\qquad\qquad\qquad\qquad$ (3.12a)

$$t_d = t_k / \gamma_{tk}$$

with the requirement:

$$t_d > t_g \qquad\qquad (3.12b)$$

where t_d is the design service life
        t_0 mean life
        t_k characteristic life (5% probability of lower value)
        γ_{t0} the central lifetime safety factor
        γ_{tk} the characteristic lifetime safety factor
        t_g target service life.

Table 3.10 The analogy between static limit state design and durability limit state design with the lifetime safety factor method

Static limit state design	Durability limit state design / lifetime safety factor method
Strength class	Service life class
Target strength	Target life
Mean strength	Mean life
Characteristic strength	Characteristic life
(5% probability of observing a lower value)	(5% probability of observing a lower value)
Design strength	Design life
Partial safety factors of materials strength	Partial safety factors of service life
Static load: mechanical	Environmental load: physical (moisture, temperature, solar radiation) chemical, biological
Partial safety factors of static loads	Partial safety factors of environmental loads
Limit states of serviceability and ultimate states	Durability limit states of serviceality and ultimate states.
Failure probability	Probability of belowing the target service life

The design formulae can then be written by applying either the performance principle or the service life principle as follows:

$$R(t_d) - S(t_d) \geq 0 \tag{3.13}$$

$$t_d - t_g > 0$$

or $\tag{3.14}$

$$t_d > t_g$$

Figure 3.13 shows a distribution of service life and the relationships between the target service life, failure probability and mean service life. The central lifetime safety factor is the relation of mean service life to the target service life.

$$\gamma_t = \frac{\mu(t_L)}{t_g} \tag{3.15}$$

where γ_t is the central lifetime safety factor,
  $\mu(t_L)$ the mean service life, and
  t_g the target service life.

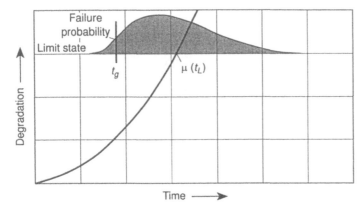

Figure 3.13 Relationship between mean service life and target service life

Using the lifetime safety factor, the requirement of target service life (corresponding to a maximum allowable failure probability) is converted to the requirement of mean or characteristic service life.

The mean or charactersitic service life is approximated by service life models which show the crossing point of the degradation curve with the limit state of durability (Figure 3.14). The mean service life evaluated by the service life model must be greater than or equal to the *design service life*, which is the product of the lifetime safety factor and target service life.

$$\mu(t_L) \geq t_d \tag{3.16}$$

$$t_d = \gamma_t t_g \tag{3.17}$$

where t_d is the design service life.

The lifetime safety factor depends on the maximum allowable failure probability. The lifetime safety factor also depends on the form of service life distribution. Figure 3.14 illustrates the meaning of lifetime safety factor when the design is done according to the performance principle. The function $R(t) - S$ is called the safety margin.

Performance behaviour can always be translated into degradation behaviour. By definition, degradation is a decrease in performance. The transformation is performed by the following substitutions:

$$R_0 - R(t) = D(t) \tag{3.18a}$$
$$R_0 - S = D_{max}$$

or

$$R_0 - R_{min} = D_{max} \tag{3.18b}$$

Figure 3.15 shows the principle of design in a degradation problem. $D(t)$ is the degrading effect of environmental loading on the performance of the structure. The range $D_{max} - D(t)$ is the safety margin.

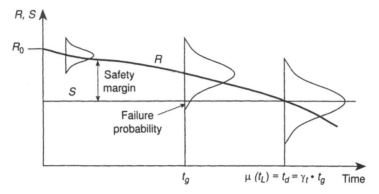

Figure 3.14   The meaning of lifetime safety factor in a performance problem

Let us consider that the degradation function is of the following form:

$$\mu(D(t)) = at^n \tag{3.19}$$

where $\mu(D(t))$ is   the mean of degradation
        a the constant coefficient
        t time, and
        n the exponent.

The exponent n may in principle vary between $-\infty$ and $+\infty$.

The coefficient a is fixed when the mean service life is known:

$$a = \frac{D_{max}}{\mu(t_L)^n} \tag{3.20}$$

Degradation is assumed to be normally distributed around the mean. It is also assumed that the standard deviation of D is proportional to the mean degradation, the coefficient of variation being constant, v_D.

Figure 3.14 shows the degradation as a function of t^n. The value of γ^t can be determined as follows:

The index β of standard normal distribution at t_g is:

$$\beta = \frac{D_{max} - D_g}{v_D D_g} = \frac{1}{v_D}\left\{\frac{D_{max}}{D_g} - 1\right\} \tag{3.21}$$

where D_{max} is   the maximum allowable degradation
        D_g the mean degradation at t_g, and
        v_D the coefficient of variation of degradation.

From Figure 3.14 we get:

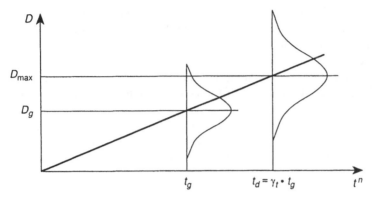

Figure 3.15 Determination of lifetime safety factors with normally distributed D

$$\frac{D_{max}}{D_g} = \frac{\left(\gamma_t \, t_g\right)^n}{\left(t_g\right)^n} = \gamma_t^{\,n} \tag{3.22}$$

By assigning this to Equation 3.21 we obtain:

$$\gamma_t = (\beta v_D + 1)^{1/n} \tag{3.23}$$

The lifetime safety factor depends on β (respective to the maximum allowable failure probability at t_g), the coefficient of variation of D and the exponent n. Thus the lifetime safety factor is not directly dependent on t_g.

DEGRADATION FACTORS AND THEIR STRUCTURAL EFFECTS

The following degradation factors are most commonly dealt with:

1 corrosion due to chloride penetration
2 corrosion due to carbonation
3 mechanical abrasion
4 salt weathering
5 surface deterioration
6 frost attack.

Additionally there exist some internal degradation processes, such as alkaline-aggregate reaction, are not treated here as they can be solved by a proper selection of raw materials and an appropriate design of concrete mix.

Degradation factors affect either the concrete or the steel or both. Usually degradation takes place on the surface zone of concrete or steel, gradually destroying the material.

The main structural effects of degradation in concrete and steel are the following:

1 Loss of concrete leading to reduced cross-sectional area of the concrete.
2 Corrosion of reinforcement leading to reduced cross-sectional area of steel
   bars. Corrosion may occur
   •   in cracks
   •   on all steel surfaces, assuming that the corrosion products are able to
        leach out through the pores of the concrete (general corrosion in wet
        conditions).
3 Splitting and spalling of the concrete cover due to general corrosion of
   reinforcement, leading to a reduced cross-sectional area of the concrete, to a
   reduced bond between concrete and reinforcement and to visual unfitness.

DESIGN PROCEDURE

The design procedure included inthis phase is presented in Figure 3.16.
   Ordinary mechanical design is performed using conventional design methods.

DURABILITY DESIGN PROCEDURE

1 specification of the target service life and design service life
2 analysis of environmental effects
3 identification of durability factors and degradation mechanisms
4 selection of a durability calculation model for each degradation mechanism
5 calculation of durability parameters using available calculation models
6 possible updating of the calculations of the ordinary mechanical design
7 transfer of the durability parameters into the final design.

DESCRIPTION OF THE PHASES OF THE DURABILITY DESIGN

Phase 1: Specification of target service life and design service life
The target service life is defined corresponding to the requirements given in
common regulations, codes and standards in addition to possible special
requirements of the client. Typical classes of service life are 10, 25, 50, 75, 100
etc. years. The safety classification of durability design is presented in Table 3.11.
   The design service life is determined by formula:

$$t_d = \gamma_t t_g \qquad\qquad (3.24)$$

where t_d is the design service life,
       γ_t the lifetime safety factor, and
       t_g the target service life.

Phase 2: Analysis of environmental effects
The analysis of environmental effects includes identification of the climatic
conditions such as temperature and moisture variations, rain, condensation of

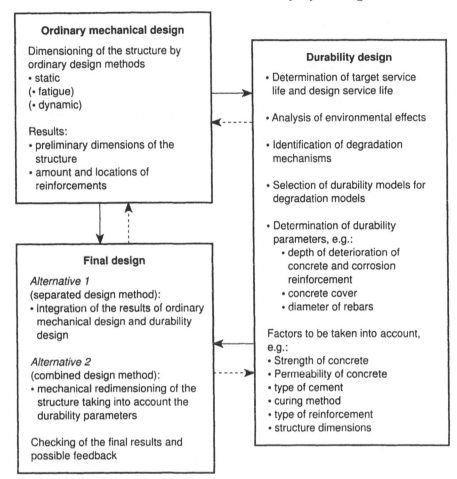

Figure 3.16 Flow chart of the durability design procedure

moisture, freezing, solar radiation and air pollution, and the identification of geological conditions such as the location of ground water, possible contact with sea water, contamination of the soil by aggressive agents like sulphates and chlorides. Man-made actions such as salting of roads, abrasion by traffic etc. must also be identified.

Phase 3: Identification of degradation factors and degradation mechanisms
Based on the environmental effect analysis the designer identifies the degradation factors to which the structure will most likely be subjected. Some kind of degradation process is usually assumed to take place in both the concrete and the reinforcement.

Table 3.11 Safety classification of durability design and the corresponding central safety factors

Limit state	Safety class of durability design	Separated design		Combined design[1]	
		Lifetime safety factor	Load and material safety factors	Lifetime safety factor	load and material safety factors
		γ_t		γ	
Ultimate limit state	1. Serious social, economic or ecological consequences of a mechanical failure.	3.3	normal[2]	2.5	normal[2]
	2. Consequences of a mechanical failure are not serious.	2.9	normal[4]	2.2	$\gamma_g 1.35$ $\gamma_p 1.38$ $\gamma_c 1.35$ $\gamma_s 1.35$
Service-ability limit state	1. Noticeable consequences and considerable repair costs	2.5	–	1.9	–
	2. Non-noticeable consequences and repair costs	1.9	–	1.5	–

Notes

1   Check for m (relative reduction of the safety margin during $0 \rightarrow t_d$) ≤ 0.7 is required.

2   Load and material safety factors specified for ordinary mechanical design are used. Reduced values of load and material safety factors may be used in the durability design. However, the safety at the start of service life ($t = 0$) must be at least the same as that required in ordinary design.

Phase 4: Selection of durability models for each degradation mechanism

A designer must determine which degradation factors are decisive for service life. The models presented in the report may be applied in these evaluations. In *concrete structures exposed to normal outdoor conditions* the effects of degradation mechanisms can be classified into the following structural deterioration mechanisms:

1   corrosion of reinforcement in cracks, causing a reduction in the cross-sectional area of steel bars
2   surface deterioration or frost attack, causing a reduction in the cross-sectional area of concrete.

Phase 5: Calculation of durability parameters through calculation models
Damage is determined using the design service life, t_d, as a value of time variable. Selected calculation models are presented in the appendix of the TC 130–CSL report (Sarja and Vesikari 1996).

Phase 6: Possible updating of calculations in ordinary mechanical design
Some durability parameters may influence the mechanical design. An increase in concrete dimensions, increases the dead load, thus increasing the load effects on both the horizontal and vertical structures.

Phase 7: Transfer of durability parameters to the final design
The parameters of the durability design are listed and transferred to the final design phase for use in the final dimensioning of the structure.

Phase 8: Final design
Separated design method   In the separated design method the mechanical design and the durability design are separated. The ordinary structural design (phase 1) produces the mechanical safety and serviceability parameters whereas the durability design (phase 2) produces the durability parameters. Both of these groups of parameters are then combined in the final design of the structure.

Combined design method   In the combined design method mechanical design is carried out taking into account the results of the durability design and the required safety at the end of service life. The combined method is especially suited to degradation mechanisms which directly affect the loadbearing capacity or the mechanical serviceability of structures. Examples of this type of design are presented by Sarja and Vesikari (1996).

3.7.3 Statistical durability design

The basic principles for statistical durability design are presented above as an introduction to the lifetime safety factor method. In the statistical method we have to know the distribution functions of environmental loading $S(t)$ and of the resistance function of the structure $R(t)$, and their possible correlation with each other. Consequently the failure probability can be calculated, and limited into values which are listed in standards as being permissible for each kind of failure or defect. Similar levels of safety as presented in the case of a lifetime safety factor method can be applied. The method is suitable for all types of structures and for all materials.

The convolution integral can be solved with modern computer programs such as 'Mathcad', or others.

The direct statistical method can be applied in special cases when the distribution functions $S(t)$ and $R(t)$ can be determined either theoretically or exprrerimentally. This can be the case, for example, at the product development stage of industrially-produced standard or modification products. In such cases, a large number of experiments are possible and the results can be applied to a large number of products.

3.8 Calculation of heat energy consumption of a building

3.8.1 Calcualtion models of heat energy consumption

Heat energy consumption can be modelled as a function of the thermal insulation factor of the building envelope, ventilation of indoor air and air leakage flow through the envelope. The thermal insulation factor includes the heat conduction through different parts of envelope including walls, roof, floor, windows and doors. Ventilation includes the heat loss with exhaust air. The following equation (3.25) can be used in calculations

$$Q_{tot} = Q_{cond} + Q_{vent} + Q_{leak} \tag{3.25}$$

where Q_{tot} is   total heat energy consumption
Q_{cond} energy loss as conduction through the envelope
Q_{vent} energy loss with ventilation exhaust air
Q_{leak} energy loss with air leakage through the envelope

The energy loss through conduction is calculated by Equations 3.26 and 3.27, energy loss with ventilation exhaust air is calculated by Equation 3.28, and the energy loss with leakage is calculated by Equation 3.29.

$$Q_{cond}\,[kWh] = (24/1000) \times \sum_{i=1}^{n}(k_i \times A_i \times S) \tag{3.26}$$

where k_i is   thermal conductivity of each part i of the envelope
            $(i=1...n)$ [W/m² °K]
A_i area of part i of the envelope $(i = 1...n)$ [m²]
S local temperature day number of the building [°d]
[°d]   the product: number of days times mean temperature of each day

The temperature day number

$$S\,[°d] = \sum_{i=1}^{N}(T_{i,in} - T_{i,out}), \quad \text{if } T_{i,in} - T_{i,out} > 7$$
$$S\,[°d] = 0 \qquad\qquad\qquad , \quad \text{if } T_{i,in} - T_{i,out} <= 7 \tag{3.27}$$

where N is   number of days in the actual time period
$T_{i,out}$ mean outdoor temperature of each day i
$T_{i,in}$ nominal mean indoor temperature = 20° C

Energy loss with ventilation exhaust air

$$Q_{vent}[kWh] = 0,008 \times V_{air, vent} \times S \times (1-r) \qquad (3.28)$$

where $V_{air, vent}$ is the volume of ventilation exhaust air [m³/h]

S local temperature day number of the building [°d] (calculate with eq. 3.27).

r efficiency coefficient of the thermal energy exchange from exhaust air $(0 <= r < 1)$

Energy loss with leakage air through the envelope is

$$Q_{leak}[kWh] = 0.008 \times V_{air, leak} \times S \qquad (3.29)$$

where $V_{air, leak}$ is the volume of leakage air [m³/h]

S local temperature day number of the building [°d]

3.8.2 Parameters

If not calculated in more detail, the following geometric parameters of buildings are used

- For all buildings:
    - relative area of windows in comparison to the floor area is 0.10
    - relative area of external doors in comparison to living floor area is 0.01
    - the height of the storey is 3 m, the room height being 2.6 m
    - the ratio of external wall area in comparison to living floor area is 0.70
    - ratio of the ground floor and roof areas in comparison to living floor area is 0.2 each.

The following thermal properties are used in different levels of energy efficiency:

1. buildings with low energy efficiency (ordinary buildings in warm climate areas and old buildings in cool climate areas):
    - walls: $k = 1.00$ W/m² °K
    - roof: $k = 0.80$ W/m² °K
    - ground floor: $k = 0.80$ W/m² °K
    - windows: $k = 4.00$ W/m² °K
    - doors: $k = 3.00$ W/m² °K
    - $V_{air, vent}$ / hm² living floor area = 0.50
    - $V_{air, leak}$/ hm² living floor area = 1.50
    - $r = 0$

2 buildings with ordinary energy efficiency (buildings in cool and cold climate areas, low energy buildings in warm climate areas):

- walls: $k = 0.30$ W/m² °K
- roof: $k = 0.25$ W/m² °K
- ground floor: $k = 0.30$ W/m² °K
- windows: $k = 2.00$ W/m² °K
- doors: $k = 0.70$ W/m² °K
- $V_{air, vent}$ / hm² living floor area = 1.25
- $V_{air, leak}$ / hm² living floor area = 0.50
- $r = 0$

3 buildings with high energy efficiency (low energy buildings in cool and cold climate areas):

- walls: k = 0.15 W/m² °K
- roof: k = 0.15 W/m² °K
- ground floor: k = 0.15 W/m² °K
- windows: k = 0.80 W/m² °K
- doors: k = 0.30 W/m² °K
- $V_{air, vent}$ / hm² living floor area = 1.25
- $V_{air, leak}$ / hm² living floor area = 0.10
- $r = 0$

3.8.3 Annual energy consumption

Annual energy consumption can be calculated with Equations 3.25 to 3.29 as follows using mean values of long term climatic statistics:

$$Q_{tot, d}(i) = Q_{cond, d}(i) + Q_{vent, d}(i) + Q_{leak, d}(i) \qquad (3.30)$$

where

$$Q_{cond, d}(i) \ [\text{kWh} / (a \ \text{m}^2 \ \text{living floor area})] = (24/1000) \times k_{int} \times S_a \qquad (3.31)$$

where k_{int} is integral thermal conductivity of the model building [W / (m² living floor area °K)]

$$S_a(°d)\sum_{i=1}^{N_a} S_i = \sum_{i=1}^{N_a}(T_{i, in} - T_{i, out}), \ \text{if } T_{i, in} - T_{i, out} > 7$$

$$S_a(°d) = 0 \qquad\qquad\qquad \text{if } T_{i, in} - T_{i, out} <= 7$$

N_a number of days in the year (i) (= 365 or 366)
$T_{i, out}$ mean outdoor temperature of each day i
$T_{i, in}$ nominal mean indoor temperature = 20°

$$Q_{\text{vent}, i}(i) \text{ [kWh} / a \text{ m}^2 \text{ living floor area)]} = 0.008 \times V_{\text{air, vent}} \times S_a \times (1 - r) \quad (3.32)$$

where $V_{\text{air, vent}}$ is the volume of ventilation exhaust air [m^3 / (h m^2 floor area)]

$$S_a(\text{°d}) \sum_{i=1}^{N_a} S_i = \sum_{i=1}^{N_a} (T_{i, \text{in}} - T_{i, \text{out}}), \quad \text{if } T_{i, \text{in}} - T_{i, \text{out}} > 7$$

$$S_a(\text{°d}) = 0 \qquad\qquad\qquad \text{if } T_{i, \text{in}} - T_{i, \text{out}} <= 7$$

N_a is number of days in the year (i) (= 365)
r efficiency coefficient of the thermal energy exchange ($0 < = r < 1$)

Energy loss with leakage air through the envelope is

$$Q_{\text{leak}, i}(i) \text{ [kWh} / a \text{ m}^2 \text{ floor area]} = 0.008 \times V_{\text{air, leak}} \times S_a \qquad (3.33)$$

where $V_{\text{air, leak}}$ is the volume of leakage air [m^3/(h m^2 floor area)]

$$S_a(\text{°d}) \sum_{i=1}^{N_a} S_i = \sum_{i=1}^{N_a} (T_{i, \text{in}} - T_{i, \text{out}}), \quad \text{if } T_{i, \text{in}} - T_{i, \text{out}} > 7$$

$$S_a(\text{°d}) = 0 \qquad\qquad\qquad \text{if } T_{i, \text{in}} - T_{i, \text{out}} <= 7$$

N_a is number of days in the year (i) (= 365)

References

Akao, Y. (1990) *Quality Function Deployment QFD: Integrating Customer Requirements into Project Design.* Productivity Press: Cambridge, MA.

American Supplier Institute (1989) *Quality Function Deployment, Awareness Annual.* American Supplier Institute, Inc: Dearborn, MI.

ASTM (1993) *Standard Practice for Measuring Life-Cycle Costs of Buildings and Building Systems.* ASTM Designation E 917–93. Werst: Conshohocken, PA.

ASTM (1995) *Standard Practice for Applying the Analytic Hierarchy Process to Multiattribute Decision Analysis of Investments Related to Buildings and Building Systems.* ASTM Designation E 1765–95.

Baumann, H. (1998) 'Life cycle assessment and decision making: theories and practice', PhD thesis. AFR report 183, Chalmers University of Technlogy, Technical Environmental Planning, Göteborg.

Bredenbals, B. and Willkomm, W. (1994) *Abfallvermeidung in der Bauproduktion.* Bauforschung für die Praxis, Bände der Reihe des IRB-Verlag, 6, TUB: Stuttgart.

Bredenbals, B. and Willkomm, W. (1996) *Neue Konstruktionsalternativen für recyclingfähige Wohngebäude.* Bauforschung für die Praxis, Bände der Reihe des IRB-Verlag, 22, TUB: Stuttgart.

CEN (European Committee for Standardisation) (1993) Eurocode 1: Basis of Design and Actions on Structures. Part 1: Basis of Design. ENV 1991-1. CEN/TC250. CEN: Brussels.

CEN (European Committee for Standardisation) (1997) Concrete: Performance, Production and Conformity. Draft prEN 202. CEN: Brussels.

CEN (European Committee for Standardisation) (1999) Performance Testing – Part 7: Performance Testing of a Mechanical Supply and Exhaust Ventilation Unit Used in a Single Dwelling. Component/Products for Residential Ventilation. CEN TC156/WG2/AH7 N5. CEN: Brussels.

European standard EN ISO 10211-1. Thermal bridges in building construction – Heat flows and surface temperatures – Part 1: General calculation methods.

European standard prEN ISO 12011-2. Thermal bridges in building construction – Calculation of heat flows and surface temperatures – Part 2: Linear thermal bridges. Final draft.

European standard prEN ISO 14683 Thermal bridges in building construction – Linear thermal transmittance – Simplified methods and default values. Final draft.

European standard prEN ISO 13788 Hygrothermal performance of building components and building elements – estimation of internal surface temperature to avoid critical surface humidity and calculation of intertitial condensation. Final draft.

Foliente, G.C., Leicester, R.H., Cole, I. and Mackenzie, C. (1998) 'Development of a reliability-based durability design method for timber construction', in Lacasse, M. (ed.) *Durability of Building Materials and Components,* volume 2, Proceedings of 8th International Conference on Durability of Building Materials and Components, 30 May–3 June, Vancouver. NRC Research Press: Ottawa.

IEA (International Energy Agency) (1998) Solar heating and colling programme. Task 23: Optimisation of solar energy use in large buildings.

ISO/DIS 15686-1. ISO TC 59/SC14. Guide for service life design of buildings. Draft standard.

ISO 6240-1980. Performance standards in building: Contents and presentation.

ISO 6241-1984. Performance standards in building: Principles for their preparation and factors to be considered.

ISO 6242. Building performance: Expression of functional requirements of users – Thermal comfort, air purity, acoustical comfort, visual comfort and energy saving in heating.

ISO 6243. Climatic data for building design.

ISO 7162-1992. Performance standards in building: Contents and format of standards for evaluation of performance.

ISO 8990-1994, Thermal insulation – Determination of steady-state thermal transmission properties – Calibrated and guarded hot box.

ISO 9699-1994. Performance standards in building: Checklist for briefing – Contents of brief for building design.

ISO 9223-1992 (F). Corrosion of metals and alloys. Corrosivity at atmospheres. Classification.

Kirk, S.J. and Dell'Isola, A.J. (1995) *Life Cycle Costing for Design Professionals.* New McGraw-Hill Inc.: New York.

Lakka, A., Laurikka, P. and Vainio, M. (1995) *Quality Function Deployment (QFD) in Construction* (in Finnish). Research Notes 1685. Technical Research Centre of Finland: Espoo.

Leicester, R.H. and Foliente, G.C. (1999) 'Models for timber decay and termite attack', in Lacasse, M. (ed.) *Durability of Building Materials and Components* volume 1. Proceedings of 8th International Conference on Durability of Building Materials and Components, 30 May–3 June, Vancouver. NRC Research Press: Ottawa.

Nofal, M. and Kumaran, M.K. (1999) 'Durability assessments of wood-frame construction using the concept of damage functions', in Lacasse, M. (ed.) *Durability of Building Materials and Components,* volume 1. Proceedings of 8th International Conference on Durability of Building Materials and Components, 30 May–3 June. Vancouver. NRC Research Press: Ottawa.

Leinonen, J. and Huovila, P. 'Requirements management in life cycle design', in Sarja, A. (ed.) *Integrated Life Cycle Design of Materials and Structures.* Proceedings of the RILEM/CIB/ISO International Symposium. RILEM Proceedings PRO 14. ILCDES 2000. RIL – Association of Finnish Civil Engineers: Helsinki.

Lindfors, L.G., Christiansen, K., Hoffman, L., Virtanen, Y., Junttila, V., Hanssen, O-J., Rønning, A., Ekvall, T. and Finnveden, G. (1995) *Nordic Guidelines on Life-Cycle Assessment,* Nord 1995:20, CE Fritzes AB: Stockholm.

Lippiatt, B. (1998a) 'Building for environmental and economic sustainability (BEES)'. CIB/RILEM Symposium: Materials and Technologies for Sustainable Construction, Gävle, June. Building and Fire Research Laboratory, National Institute of Standards and Technology: Gathersburg, MD.

Lippiat, B. (1998b) *BEES 1.0. Building for Environmental and Economic Sustainability. Technical Manual and User Guide.* NISTIR 6144. NIST, Building and Fire Research Laboratory, National Institute of Standards and Technology, Gathersburg, MD.

Nieminen, J. and Huovila, P. (2000) 'Quality function deployment (QFD) in design process decision-making', in Sarja, A. (ed.) *Integrated Life Cycle Design of Materials and Structures.* Proceedings of the RILEM/CIB/ISO International Symposium. RILEM Proceedings PRO 14, ILCDES 2000. RIL Association of Finnish Civil Engineers: Helsinki.

Norberg-Bohm, V., Clark, W.C., Bakshi, B., Berkenkamp, J., Bishko, S.A., Koehler, M.D., Marrs, J.A., Nielsen, C.P. and Sagar, A. (1992) *International Comparisons of Environmental Hazards: Development and Evaluation of a Method for Linking Environmental Data with the Strategic Debate Management Priorities for Risk Management.* CSIA Disc. Paper 92–09. Center for Science and International Affairs: Cambridge, MA.

Norris, G.A. and Marshall, H.E. (1995) *Multiattribute Decision Analysis Method for Evaluating Buildings and Building Systems.* NISIR 5663. National Institute of Technology: Gathersburg, MD.

Radünz, A. (1998) *Bauprodukte und gebäudebedingte Erkrankungen.* Springer Verlag: Berlin, Heidelberg.

Roozenburg, N. and Eekels, J. (1990) *EVAD, Evaluation and Decision in Design. (Bewerten und Entscheiden beim Konstruiren).* Schriftenreihe WDK 17, Edition HEURISTA: Zürich.

Sarja, A. (1989) *Principles and Solutions of the New System Building Technology (TAT).* Research Report 662. Technical Research Centre of Finland: Espoo.

Sarja, A. (1995) 'Methods and methodology for the environmental design of structures', RILEM Workshop on Environmental Aspects of Building Materials and Structures. Technical Research Centre of Finland: Espoo.

Sarja, A. (1997b) 'Framework and methods of life cycle design of buildings', Symposium: Recovery, Recycling, Reintegration, R'97, 4–7 February, Geneva. *EMPA,* VI, 100–105.

Sarja, A. (1999a) 'Environmental design methods in materials and structural engineering'. *Materials and Structures* 32 (December) 699–707.

Sarja, A. (1999b) 'Towards life cycle oriented structural engineering', in Eligehausen, R. (ed.) *Construction Material: Theory and Application.* Ibiidem-Verlag: Stuttgart.

Sarja, A. (2000) 'Durability design of concrete structures. Committee Report 130–CSL. *Materials and Structures/Matériaux et Constructions* 33 (January–February), 14–20.

Sarja, A. (2001) *Lifetime Structural Engineering* (in Finnish). Guidelines RIL 216–2001. Finnish Association of Civil Engineers: Helsinki.

Sarja, A. and Hannus, M. (1995) *Modular Systematics for the Industrialised Building.* VTT Publications 238, Technical Research Centre of Finland: Espoo.

Sarja, A. and Lautanala, M. (eds) (1993) *Future Buildings, Innovative Low Energy Concepts. IEA: Energy Conservation in Buildings and Community Systems.* VTT Symposium 139. VTT Technical Research Centre of Finland: Espoo.

Sarja, A. and Vesikari, E. (1996) *Durability Design of Concrete Structures.* RILEM Report Series 14. E&FN Spon: London.

Udo de Haes, H.A., Jolliet, O., Finnveden, G., Hauschild, M., Krewitt, W. and Müller-Wenk, R. (1999). 'Best available practice regarding impact categories and category indicators in life cycle impact assessment', Background Document for the Second Working Group on Life Cycle Impact Assessment of SETAC-Europe (WIA-2). *Int. J. LCA* 4(2): 66–74 and 4(3): 167–174.

United States Environmental Protection Agency (1990) *Reducing Risk: Setting Priorities and Strategies for Environmental Protection.* SAB-EC-90–021. United States Environmental Protection Agency, Science Advisory Board: Washington, DC.

VDI (1997) VDI-Richtlinie: VDI 2222, Blatt 1, Konstruktionsmethodik, Methodisches Entwickeln von Lösungsprinzipien.

VDI (2000) VDI-Richtlinie: VDI 2243, Blatt 1, Recyclingorientierte Produktentwicklung.

Zairi, M. and Youssef, M.A. (1995) 'Quality function deployment: a main pillar for successful total quality management and product development'. *International Journal of Quality and Reliability Management* 12(6), 9–23.

4 Design for recycling using concrete and masonry as an example

Christoph Müller

4.1 Introduction

The ecological view regarding building materials, building products and buildings is gaining more and more importance. This takes into account the technical, economic and ecological specifications for building materials, building products and buildings during the total life cycle (production, use, demolition and reuse). Considering the universal material and energy consumption since the beginning of the industrial era, the aims here, as in all other parts of life, are increasing the fixed value of products (through conservation of natural resources as well as avoiding waste) while simultaneously decreasing the effects on the environment. Examples of environmental influences are the consumption of resources, the consumption of non-renewable energy, as well as emissions into water, soil and the air.

The multiple use of substances or products is one way of conserving natural resources (and energy) and avoiding waste. This process is generally termed 'recycling' and it describes the reuse or exploitation of products or parts of products in cycles.

Material cycles, in the real sense of recycling, have only been achieved for steel. The first steps have been taken for mineral building materials. Processed and partly-classified mineral building materials (concrete and masonry rubble) are not part of a material cycle, and are used almost exclusively in road construction or as bedding materials (secondary usage). This also applies to cement-based building materials (concrete etc.) as well as masonry.

Human beings have always built living and working spaces as economically as possible and made use of the construction properties of locally-available building materials. But the demands for comfortable living conditions have risen greatly compared with the past; living, heating and ventilation habits have changed. This has led to higher demands on modern construction elements, demands which cannot be met with traditional construction materials.

In order to satisfy today's demands, various materials are used, either separately or in combination. The properties and combinations of these materials have been optimised with regard to technical requirements (requirements relating to exposure, heat insulation, sound insulation), but not with regard their subsequent reuse.

Mixing materials together often presents a hindrance when trying to bring them back into material cycles so as to enable the recycling of building materials, because these mixtures are partly incompatible in the recycling process. Separating material combinations for reuse purposes in 'real' material cycles is not possible in most cases, or feasible only with disproportionate expenditure (high energy input).

4.2 Principles for design for reuse and recycling

When designing for reuse and recycling, one can differentiate between:

- short-term measures to use building materials and buiding material combinations currently available, in order to design for reuse and recycling as far as it is possible today; and
- long-term strategies to optimise building materials, building materials combinations and building parts in terms of their ability to be recycled.

Basic strategies regarding the choice and the combination of building materials can already be determined and used today. According to the general definition of recycling, some basic rules were laid down in 'Design methods for reuse and recycling of technical products' by the Association of German Engineers, to give engineers guidelines when designing technical products 'for reuse and recycling' (VDI 1993). Although these rules apply to the metal and plastics industries, and not civil engineering, they are nevertheless applicable.

Apart from the requirement of minimisation of waste during production (building construction = low-waste site), the following three basic rules are applicable to construction principles in structural engineering:

- reduction of the variety of materials,
- avoiding insoluble composite substances and/or composite substances that are either only slightly soluble or soluble only within a high expenditure (energy input),
- separation in construction of building parts and materials/substances with different service lives and different recycling techniques.

These strategies can be applied to the field of civil engineering with respect to building construction by defining the following requirements (Bredenablas and Wilkomm 1995):

- compatibility of materials for joint processing (material recycling),
- ability to separate materials which cannot be recycled together,
- use of demountable structural components.

These specifications can be used generally when considering the separation of building materials in loadbearing frames and the interior work (including building services installations).

Table 4.1 Construction planning and building stages and their influence on the avoidance and reuse of wastes (modified from Wilkomm 1990)

Building construction	Avoidance of wastes			Reuse and recycling
	Fixed value through sustainable construction	Low-waste construction and site	Consideration of reuse and recycling	Use of secondary raw materials, recycled materials and construction components
1	2	3	4	5
Client's specifications	✔	✓	✔	✔
Design of the load-bearing structure	✔	✔	✔	✓
Design of the finishing details	✔	✔	✔	✔
Specification	✔	✔	✓	✔
Workmanship, supervision	✔	✔	–	–

✓ Some influence on the avoidance and the reuse of wastes
✔ Particular influence on the avoidance and the reuse of wastes
– Little or no influence on the avoidance and the reuse of wastes
▒ Design for reuse and recycling

If the above-mentioned specifications are to be applied in detail to specific construction components of the building shell (which, for almost all buildings, mainly consists of mineral building materials), the materials cycles in which the reuse or further use can/must take place have to be specifically defined. The material properties of the construction components have to be determined based on assumptions concerning their ability to be separated and processed. The ability of building materials and construction components to be recycled depends on the degree and/ or technical level of the desired reuse. This evaluation can be included in building materials and construction component catalogues, which can be made available to all those involved in building so as to exert a direct or indirect influence on the design for reuse and recycling of the planned construction. Table 4.1 shows which stages of the design process can influence the avoidance and the reuse of wastes, especially through the active process of a design for recycling.

In the long run, a 'one dimensional recycling plan' (a material cycle in one area of use) for mineral building materials is not very helpful. Generally, the application of various cycles and the optimisation of the material use can lead to a distinct reduction of wastes. Figure 4.1 shows a scheme for a connected materials cycle system using the example of cement-based and lime-based building materials.

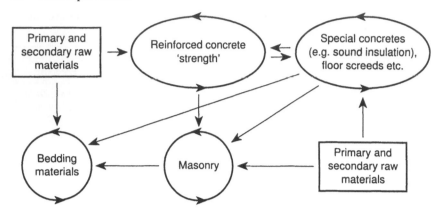

Figure 4.1 Connected materials cycle based on the example of cement- and lime-based
building materials

The primary aim is to keep building materials or construction components in their own materials cycle for as long as possible (direct usage) and to minimise the amounts with a low specification level (downcycling) that is introduced into another cycle. Another essential aim is to control the use of all mineral building materials when building, so that reuse (according to the basic rules mentioned above) is possible. These basic rules have to be considered when evaluating new buildings and future developments of building materials and building material combinations for their ability to be recycled.

4.3 Examples of building materials recycling

General

A survey of current building materials and construction techniques and the ability of the building materials to be reused has to be undertaken before the recyclability of mineral building materials and building parts can be optimised. This begins with the details of building materials and is then applied to construction components. The following list of state-of-the-art recycling possibilities for mineral building materials is based, apart from concrete, exclusively on German experience. These results will be complemented with the results from international research projects.

Concrete and recycled aggregates for concrete in building construction

The evaluation of numerous experiments (for example Forster *et al* 1994, Hansen 1992, Wesche and Schulz 1982) has shown that, in most cases, pure demolished concrete as well as demolished masonry can basically be used as aggregates in the production of new concrete.

A proposal for the characterisation of concrete with recycled aggregates, on the basis of the results in Hansen (1992) can be found in the 'Specification for Concrete with Recycled Aggregates' (RILEM 1994). It differentiates between three categories of recycled coarse aggregates (> 4 mm).

* Type I: Aggregates which are implicitly understood to originate from masonry rubble.
* Type II: Aggregates which are implicitly understood to originate from concrete rubble.
* Type III: Aggregates which are implicitly understood to consist of a blend of recycled aggregates and natural aggregates.

Recycled aggregates of type III have to contain at least 80% by mass of natural aggregates and may contain only up to 10 % by mass of aggregate type I.

Aggregates are assigned to the three categories according to the criteria presented in Table 4.2. The areas of use of recycled aggregates of types I–III is based on the exposure classes in DIN V ENV 206 (1990) and, depending on the expected application, requires the corresponding concrete tests shown in Table 4.3.

The properties of concretes produced with these recycled aggregates can differ (compared with concretes of identical composition but produced with natural aggregates), depending on the amount of recycled aggregates used (i.e. ratio of recycled aggregate to total aggregate volume). Because of these property changes, corresponding factors were catered for in RILEM (1994) and these factors assume a 100% exchange of natural aggregates for recycled aggregates (worst case scenario). These figures show that in order to maintain the same technical specification profiles of concrete as a building material, the proportion from its own material cycle, as well as the proportion of secondary raw materials introduced into the cycle from other areas has to be limited if these property changes are not compensated for by other measures (for example,. lower water cement ratio w/c higher cement strength class etc.). Because of this, the direct exploitation quotient of the concrete inevitably drops during each recycling cycle.

Clay bricks and clay brick masonry

The reuse of clay brick material in the production of new bricks is possible only in relatively small amounts. Pure type material (without residual mortar) can be added to the raw material mixture in the region of 10 to 20% by mass in the form of ground brick chips as a filler. Greater amounts can lead to losses in strength. The question whether residual mortar or other residual materials influence brick production or brick properties has not yet been determined. At present, the German clay brick industry does not see any urgent need to carry out research because of present market conditions (Schubert and Heer 1997)

The use of clay brick masonry rubble is possible as a concrete aggregate and for various applications in road construction. Thus, at the moment, direct use in the 'clay brick' cycle is minimal for pure type material and not possible for that with residual mortar attached.

Table 4.2 Classification of recycled coarse aggregates for concrete (RILEM 1994)

Specification		Unit	Type I	Type II	Type III
Dry particle density		kg/m³	≥ 1500	≥ 2000	≥ 2400
Water absorption			≤ 20	≤ 10	≤ 3
Materials	< 2200kg/m³			≤ 10	≤ 10
with	< 1800 kg/m³		≤ 10	≤ 1	≤ 1
SSD[1]	< 1000 kg/m³ [2]		≤ 1	≤ 0.5	≤ 0.5
Foreign materials (metal, glass, soft components, bitumen) content		% by mass	≤ 5	≤ 1	≤ 1
Metal content			≤ 1	≤ 1	≤ 1
Organic material content			≤ 1	≤ 0.5	≤ 0.5
Materials < 0.063 mm contnent			≤ 3	≤ 2	≤ 2
Sand (< 4 mm) content			≤ 5	≤ 5	≤ 5
Sulphate (SO₃) content			≤ 1	≤ 1	≤ 1

Notes

1   Water saturated surface dry condition
2   % by mass and % by volume

Lightweight concrete blocks and lightweight concrete masonry

Specific processing of mainly pure type pumice demolition rubble is not carried out in Germany. Some companies in the pumice processing industry have started to collect their production excess (called recycled pumice) and prepare it as a pumice replacement. Recycled pumice is used to make up to roughly 10% by volume of the aggregate content in (lightweight) concrete blocks of density classes greater than 0.8 (Tebbe and Hoffmann 1995). Tebbe and Hoffman (1995) examined the possibility of substituting pumice with recycled pumice from production excesses as well as pumice demolition rubble in the production of lightweight concrete for lightweight concrete blocks. Based on this data, recycled pumice can replace natural pumice by up to 20% by volume when producing lightweight concrete blocks of density class 0.9, if the bulk density of the recycled pumice is £ 0.60 kg/dm³ and/or the grain density is £ 0.95 kg/dm³. When producing light-weight concrete blocks of density class 1.0, up to 5% by volume of natural pumice can be replaced by pumice demolition rubble, if the bulk density of the pumice demolition rubble is £ 0.86 kg/dm³ and/or the grain density is £ 1.36 kg/dm³ and the proportion of 'non-pumice building material' does not exceed around 46% by mass (the upper limit of the region investigated by Tebbe and Hoffmann).

Sand-lime bricks and sand-lime brick masonry

The principal opportunities when producing sand-lime bricks for reusing defective blocks from the production process, as well as bricks from the demolition of buildings were investigated by Eden (1995, 1997).

Table 4.3 Provisions for the use of concrete with recycled aggregates (RILEM 1994)

Exposure class	Recycled aggregate		
	Type I	*Type II*	*Type III*
	Maximum strength class		
	$C16/20^{4,5}$	$C\ 50/60$	*no limit*
	additional testing		
1 Dry	none		
2a Dry without frost	• ASR expansion test[1]	• ASR expansion test[1]	
4a Seawater environment without frost			
2b Moist with frost	Not permitted	• ASR expansion test[1] • Bulk freeze thaw test[2]	
4b Seawater environment with frost			
3 Moist with frost and de-icing salt		• ASR expansion test[1] • Bulk freeze thaw test[2] • De-icing salt test[3]	

Notes

1 Expansion test to evaluate alkali silica reactivity according to national regulations
2 ASTM C666
3 SS 137244
4 Conforming with ENV 206

The results obtained by these investigations can be summarised as follows:

• The reuse of absolute pure type sand-lime brick demolition material as an addition to the raw mixture during sand-lime brick production is possible without any seriously detrimental effects on the properties of the bricks. This also applies equally to a 100% exchange of the grain fractions > 1.0 mm (proportion of sand-lime demolition material approximately 21–23% by mass of the raw mixture), ≤ 0.2 mm (proportion of the sand-lime demolition material is approximately 7% by mass of the raw mixture) and ≤ 0.2 up to 1.0 mm (proportion of sand-lime demolition material approximately 60% by mass of the raw mixture).

• The reuse of sand-lime brick demolition material with mortar residue (proportion of mortar residue up to approximately 11% by mass) as an addition to the raw mixture during the production of sand-lime bricks is possible without

Table 4.4 Factors in the evaluation of the material properties of concrete with recycled aggregates (RILEM 1994)

Parameter	Recycled aggregate		
	Type I	*Type II*	*Type III*
Tensile strength f_{ctm}	0.85	1.00	1.00
Modulus of elasticity E_{cm}	0.65	0.80	1.00
Creep coefficient $(\varphi\infty, \varphi_{t0})$	1.00	1.00	1.00
Shrinkage (εc_{s0})	2.00	1.50	1.00

deleterious effects on the properties of the sand-lime bricks. This also applies in the same way to an exchange of the grain fractions > 1.0 mm (proportion of sand-lime demolition material approximately 20% by mass of the raw mixture), £ 0.2 mm (proportion of sand-lime demolition material approximately 6% by mass of the raw mixture) and £ 0.2 up to 1.0 mm (proportion of sand-lime demolition material approximately 60% by mass of the raw mixture).

•   The addition of sand-lime brick demolition material with foreign substances (among others, thermal insulation composite systems based on polystyrol or mineral wool and components from bitumen seal sheets were investigated) leads to high losses in the quality properties of the bricks (for example, reduction in compressive strength) compared with bricks produced under the same conditions but without sand-lime demolition material.

The application of this knowledge to production conditions in a sand-lime brick plant depends on the specific situation (type of sand, type of sand-lime bricks, machine equipment) which needs to be checked by pre-testing. For future recycling of sand-lime bricks, it is recommended that a pure type separation of the demolition material is carried out in order to fulfil requirements for using demolished sand-lime masonry in the production of sand-lime bricks (Eden 1997).

Aerated concrete and aerated concrete masonry

In Germany, the collection of pure type aerated concrete waste for processing in a plant is guaranteed by the aerated concrete industry through the DIN plus certification programme (DINCERTCO 1997). According to this programme, pure type means that only small amounts of mineral plaster and mortar as well as (for aerated concrete building parts) the usual layers in aerated concrete demolition may be present. Nevertheless, the proportion of these materials that are technically harmless is not quantified. It is currently possible to add aerated concrete rubble with low residual amounts of mineral plaster and mortar as well as for the customary layers of aerated concrete building parts up to 10% by mass of the dry fomulation without loss of quality (Schober 1997 cited in Schubert and Heer 1997). Fine aerated concrete demolition material is nevertheless mainly used as heat insulation or as a substrate for green roofs (Weber and Hullmamm 1991, 1995).

Mortar

The reuse of mortar is only practically achievable during production of factory-mixed mortars. In Germany, dry production wastes resulting from this are completely reused in the production process (Gänßmantel 1997).

The separation of mortar and bricks/blocks during demolition of concrete or masonry constructions requires a lot of effort since a good bonding between mortar and brick is naturally an important criterion for the quality of the mortar. Since the mortar, in comparison to bricks, represents only approximately 3% by mass of normal masonry constructions, the effect of plaster and mortar on the ability of concrete or bricks to be recycled is of much greater importance than vice versa. In this connection, the possibility of recycling residual substances from other cycles during mortar production is of interest. The possibility of processing six different recycled fine aggregates from prepared, pure type and mainly lime and/or lime-containing building materials was investigated by Hoppe *et al.* (1994). It was discovered that recycled fine aggregates seemed to be suitable primarily in the production of normal mortar. Materials which tend to be more like lightweight aggregates, such as lightweight clay brick sand and aerated concrete sand, were found to be more unsuitable for this sort of use than 'dense' materials from normal concrete, sand-lime bricks or mortar.

Plaster

The reuse of plaster has the same difficulties as the reuse of mortar. It can only be reused in combination with concrete and/or masonry, because separation is not possible in most cases or feasible only with disproportionate expenditure (Schubert and Herr 1997). With regard to design for reuse and recycling it has to be determined whether the reuse and recycling of concrete and masonry is affected by plaster (especially gypsum plaster and polymer modified plaster), consequently the development of reused types of plasters is of special interest.

Comparisons of different construction alternatives

If design engineers are to be able to design for reuse and recycling, they need tools to compare different alternatives for construction elements (for example, walls or ceilings). A catalogue of different types of walls, with comparisons of their reuse capabilities, could be such a tool. In general, such an evaluation can be done for any kind of wall, but for comparisons it is necessary to review different construction alternatives which meet the same level of technical requirement.

Such schedules have to be based on the current state of knowledge and will need to be revised at appropriate intervals. Besides containing information on the chemical composition (including environmentally relevant substances) and the energy required in production processes, these schedules should contain guidelines for the future reuse and recycling. An analysis of various construction alternatives, as in the example shown in Figure 4.2 can be easily differentiated if reuse considerations are included in the decision-making process. This assumes that reuse is 'basically possible', but can also include the determination (as accurately as possible) of substance parameters of a construction as regards their reuse in comparison to the field of application. In the following example, the technical specifications for an outer wall were defined based on the coefficient of heat transfer k. An example is given for walls with k of approximately 0.44 W/(m²K). The different load-bearing capacities and economic feasibilities have to be noted at this point.

The required level of performance is decisive for such evaluations. In the examples the representation of various aspects in the assessment of the recycling capability is based on the background of the possibilities, for the usability of the substances and the allocation of mineral building material mixtures used as concrete aggregate (RILEM 1994) as well as in road construction (BRB 1996).

For composite substances and/or for walls with more than one layer, the use of these main components within their own cycles would result in a different evaluation than if utilisation on all levels were acceptable and only substances that are either hard to or impossible to reuse when the composition disintegrates are excluded. This is the case if the material cycle is defined as that of the building material which represents the greatest proportion of the mass. This classification is based on the need to keep every material in its own cycle as long as possible to avoid wastes. What all four examples of the walls have in common is that they consist solely of mineral building materials and thus satisfy the basic requirement of minimising material variety. The avoidance of gypsum plaster (interior plaster) and/or heat insulating plaster with non-mineral aggregates is a further contribution to the optimisation of the recycling capability.

While Wall 1 (single-layered brick outer wall) has a load-bearing function as well as being largely responsible for total heat conservation, Walls 2–4 fulfill these functions with two wall layers and/or two wall layers plus additional insulation which is either totally separated or through the construction joint. In the case of Walls 3 and 4, mineral fibre is a hindrance during material preparation; separation of the mineral fibre boards from the rest of the construction is basically possible, but only with additional effort. With complete material utilisation ignoring the level of reutilisation as an aim, then Walls 1 and 2 are the most appropriate as they can be used, without having to be separated, for road construction. However, reutilisation at the same technical level is only possible to a small extent or not possible at all. In the case of Wall 3, the mineral components can be used to produce new bricks (that is, in their own cycle), when separated from insulation materials during demolition. A theoretical possibility of use as a concrete aggregate and in road construction also exists.

Wall 4 offers the possibility of using part of the facade (the facing concrete) in element recycling (i.e. without further effort spent on material preparation). The loadbearing concrete wall can be recycled and reused in concrete production when separated from the insulation material during demolition. Separating the concrete from the steel reinforcement does not present any problems and reusing the steel

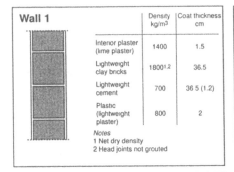

Wall 1

	Density kg/m3	Coat thickness cm
Interior plaster (lime plaster)	1400	1.5
Lightweight clay bricks	1800[1,2]	36.5
Lightweight cement	700	36 5 (1.2)
Plastic (lightweight plaster)	800	2

Notes
1 Net dry density
2 Head joints not grouted

Wall 2

	Density kg/m3	Coat thickness cm
Interior plaster (lime plaster)	1400	1.5
Lightweight concrete blocks	1000[1,2]	24.0
Lightweight cement	1000	24.0 (1.2)
Heat insulation plaster[3]	450	6.0

Notes
1 Lightweight concrete with natural pumice
2 Head joints not grouted
3 Mineral aggregates

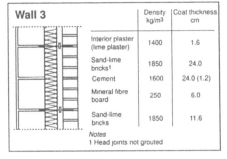

Wall 3

	Density kg/m3	Coat thickness cm
Interior plaster (lime plaster)	1400	1.6
Sand-lime bricks[1]	1850	24.0
Cement	1600	24.0 (1.2)
Mineral fibre board	250	6.0
Sand-lime bricks	1850	11.6

Notes
1 Head joints not grouted

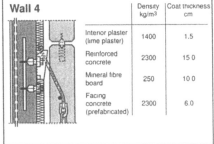

Wall 4

	Density kg/m3	Coat thickness cm
Interior plaster (lime plaster)	1400	1.5
Reinforced concrete	2300	15 0
Mineral fibre board	250	10 0
Facing concrete (prefabricated)	2300	6.0

Figure 4.2 Examples for walls with a coefficient of heat transfer $k \approx 0.44$ W/(m²K)

reinforcement is possible. If the insulation is not separated during the demolition phase, the effort needed in material preparation increases, because apart from grinding and sieving, a further wet or dry preparation is necessary. The interior plaster will be inserted mainly in the sand fraction of the demolished concrete.

The four examples given here show that the question of the recycling capability of parts of buildings, even when they consist solely of mineral building materials, cannot be answered with a definite 'yes' or 'no', and the assessment of the recycling capability of mineral building materials and building material combinations is currently hard to quantify. Even when non-mineral components are avoided as far as possible, the assessment might not be uniform, depending on the level of reuse and recycling that is to be attained. Furthermore, it should be considered that from a general ecological point of view, the additional energy needed for further use (compared to the use of a corresponding primary raw material) and the corresponding emissions, can certainly be a criterion for excluding specific utilisation options.

4.4 Reuse of structural components

To achieve buildings that can be reused and recycled, construction elements with different materials cycles which cannot be recycled together, must be demountable. Demountable construction elements can be realised in all parts of building structures:

Wall 1

+ Purely mineral
− Separation plaster/mortar and brick
  not possible or only partly possible by
  crushing
− 'Non-brick' proportion 18% mass
  → permissible amount of 'non-brick'
    components for brick production
    not known
+ Aggregate for lightweight concrete
− Concrete aggregate: Dry density
  $\rho \approx 1420$ kg/m³
  → not applicable according to RILEM
− Shape factor?
+ Filling of utility and duct trenches,
  substructures, noise abatement walls,
  parks

Wall 2

+ Purely mineral
− Separation plaster/mortar and brick
  not possible or only partly possible by
  crushing
− 'Non-lightweight' concrete proportion
  24.6% by mass
  → Dry density $\rho \approx 900$ kg/m³
    applicable in the production of
    lightweight concrete blocks
    of the density class 1.0
− Concrete aggregate:
  Dry density $\rho \approx 900$ kg/m³
  → not applicable according to RILEM
+ Filling of utility and duct trenches,
  substructures, noise abatement walls,
  parks

Wall 3

+/− Purely mineral/mineral fibre is
  foreign material in the production of
  sand-lime bricks
+ Facing masonry can be demolished
  separately
+ Demountable anchorage for heat
  insulation is available
− Recycling for mineral fibres?
  → possible reuse of mineral
    fibre boards
+ Amount of plaster and mortar
  backup cement ≈ 9% by mass
  → applicable in the production
    of sand-lime bricks
− Amount of cement in facing
  masonry ≈ 17% by mass
+ Dry density of mixture of backup
  masonry $\rho = 1770$ kg/m³
  → concrete aggregate (theoretical)
+ Utility and duct trenches,
  substructures, noise abatement
  walls, parks

Wall 4

+/− Purely mineral/mineral fibre is
  foreign material as a concrete
  aggregate
+ Facade elements can be reused
  separately
+ Demountable anchorage for heat
  insulation is available
− Recycling for mineral fibres?
  → possible reuse of mineral
    fibre boards
+ Dry density of mixture of
  concrete/plaster $\rho = 2218$ kg/m³
  → applicable as concrete
    aggregate according to RILEM
− Materials with dry density < 1800
  kg/m³ 5.7% by mass > 1.0% by
  mass
  → enrichment of the sand
    proportion probable
+ Base courses, utility and duct
  trenches, substructures, noise
  abatement walls, parks

Figure 4.3 Classification of the separability and the recycling capability of the walls shown
in Figure 4.2

- Building shell
  - Separation of load-bearing and non load-bearing elements,
  - No monolithic connections (if possible),
  - Use of prefabricated buildings with demountable connections,
  - Demountable connections between building shell and interior work.
- Interior work
  - Sandwich arrangements for walls,
  - No composite constructions (if possible),
  - Separation of exposed and non-exposed building components,
  - Separation of building components with and without heat insulation requirements,
  - Separation of building components with and without fire protection requirements.
- Installations
  - Accessibility and interchangeability of installations,
  - Use of service block modules,
  - No flush-mounted installations (if possible).

Not every building is suitable for demountable construction. The use of demountable construction is especially appropriate for industrial buildings, halls, or modular building systems for schools and housing.

Examples of prefabricated demountable constructions exist in the Netherlands and Austria (Reinhardt 1976, 1985, Reinhardt *et al.* 1984), demountable roof frames and columns where used for the building of the Munich airport (Graubner and Reiche 1997, Walraven 1988).

Many aspects have to be taken into consideration in the development of demountable und reusable prefabricated sections. Prefabricated sections should be as large as possible (provided they can be transported) so they are quick and easy to assemble or dismantle. On the other hand, demountable building elements should be relatively small and standardised, so they can be used in a wide range of new applications. The connection elements are a very important part of demountable.

Recycling of building elements certainly presents a basic possibility for design and construction for recycling, but nevertheless, because of logistics problems (such as the storage of the elements to be reused) and aesthetic considerations (as in case of the German *Plattenbauweise* precast panel building system of the 1960s and 1970s) there is only a very small chance of it being realised in a large number of applications.

4.1.5 *Conclusions*

Changes in human living habits has led to the development of building methods using a wide variety of materials. In addition to the consumption of raw materials, consumption of energy in the production of building materials (and production of associated emissions), there are also negative environmental effects caused by the accumulation of building waste after demolition. These negative environmental effects cannot be reduced to 'zero'. In future, the responsibility of the design engineers should not end with the handing over of buildings to the client, but should include the whole life cycle up to demolition and reuse with respect to recycling. This also entails the selection of building materials and construction methods for future reuse or recycling. In order to achieve this, the designer must have the necessary tools at his disposal for example, in the form of catalogues from which to compare and select building materials and building elements.

Essential considerations when selecting building materials are:

• the use of environmentally compatible secondary raw materials and recycled building materials
• the reusability (design for recycling) of building materials at various utilisation levels.

Until now, many mineral building materials could only be reused on a low technical level. Evaluation of current reuse possibilities for building materials within the same cycle shows that relatively low utilisation levels can be achieved, even in pure type cases. The combination of various material cycles is especially promising, but should nevertheless take account of the requirements of the utilisation options. For this it is especially necessary to develop

• new optimally recyclable building materials (while considering recycling materials and secondary raw materials) including their respective recycling techniques, as well as
• compatible composite constructions or
• easily separable material compositions (to overcome incompatibility).

The reuse of building elements is recognised as a possibility, although it has only very small chances of being realised because of the logistical problems and aesthetic considerations. In addition to this, demolition of pure type reinforced concrete constructions (skeleton structures) is cheap and does not present any problems. Problems and high costs always originate with inseparable 'mixed demolition waste'.

Finally, the possibilities for varied usage of buildings through flexible building design should be increasingly considered.

References

BRB (Bundesverband der Deutschen Recycling-Baustoff-Industrie e. V.) (1996) *Richtlinie Recyclingbaustoffe. Eigenschaften, Anforderungen, Prüfungen und Überwachung.* Bundesverband der Deutschen Recycling-Baustoff-Industrie e. V. (BRB).

Bredenbals, B. and Willkomm, W. (1995) *New Construction Alternatives for Reuse and Recycling in Residential Construction.* Best.-Nr. F 2280, Institut für Industriealisierung des Bauens, Hannover (Hrsg.), Forschungsbericht Nr. BI5 80 01 93–28. IRB Verlag: Stuttgart.

DIN V ENV 206 Beton; Eigenschaften, Herstellung, Verarbeitung und Gütenachweis. Ausgabe Oktober 1990.

DINCERTCO Gesellschaft für Konformitätsbewertung (1997) *Zertifizierungsprogramm: Richtlinien für die Erteilung einer Genehmigung zum Führen des DIN plus-Zeichens. Teil B: Besondere Bestimmungen für Porenbetonprodukte nach DIN 4165, DIN 4166 und DIN 4223.* DINCERTCO Gesellschaft für Konformitätsbewertung mbH: Berlin.

Eden, W. (1995) 'Wiederverwertung von Kalksandsteinen aus Abbruch von Bauwerken bzw. aus fehlerhaften Steinen aus dem Produktionsprozeß', *Kurzberichte aus der Bauforschung* 36(4), 197–198

Eden, W. (1997) *Herstellung von Kalksandsteinen aus Bruchmaterial von Kalksandstein-mauerwerk mit anhaftenden Resten von Dämmstoffen sowie weiterer Baureststoffe.* Forschungsbericht Nr. 9978. Forschungsvereinigung Kalk-Sand e.V.: Hannover.

Forster, S.W., Moore, S.P. and Simon, M.J. (1994) 'Behavior of recycled concrete as aggregate in concrete', in Malhotra, V.M. (ed.) *Third CANMET ACI National Conference on Durability of Concrete*, 22–28 May, Nice. CANMET: Ottawa.

Gänßmantel, J. (1997) 'Ökologische Aspekte von Putz- und Mauermörtel: Rohstoffe, Herstellung, Transport und Verarbeitung', in *13. International Baustofftagung, – ibausil* 24–26 September, Weimar. Bauhaus-Universität: Weimar.

Graubner, C.-A. and Reiche, K. (1997) 'Recyclinggerechtes Bauen mit demontierbaren Betontragwerken', *Darmstädter Massivbau-Seminar* 18, VI 1–22. Freunde des Instituts für Massivbau: Darmstadt.

Hansen, T.C. (1992) RILEM: *Recycling of Demolished Concrete and Masonry.* Report of Technical Committee 37–DRC: Demolition and Reuse of Concrete. Chapman & Hall: London.

Hoppe, B., Mehlmann, M. and Kazemi, S. (1994) *Verwertbartbarkeit und Umwelt-verträglichkeit kalkhaltiger Recyclingmaterialien.* Forschungsbericht Nr. 3/94 / M 007 i. Forschungsgemeinschaft Kalk und Mörtel e.V.: Köln.

Menkhoff, H. and Deters, K. (1992) *Recycling von Wandbaustoffen im Hochbau.* Forschungsbericht Nr. G 469. Institut für Bauforschung: Hannover.

Reinhardt, H.W. (1976) *Demontable Betongebäude?* Technische Hochschule: Delft.

Reinhardt, H.W. (1985) 'Demontabel bauen mit Beton', *Betonwerk und Fertigteil-Technik,* 51(5), 300–305

Reinhardt, H.W., Kolpa, J.J. and Stroband, J. (1984) 'Bauen mit demontablen Beton-fertigteilen in den Niederlanden, Teile 1+2', *Betonwerk und Fertigteil-Technik* 50(1), 22–27, 50(2), 105–110.

RILEM (1994) 'Specification for concrete with recycled aggregates', RILEM Recommendation: 121–DRG Guidance for demolition and reuse of concrete and masonry, *Materials and Structures* 27, 557–559.

Schober, Dr (Hebel AG, Fürstenfeldbruck) (1997) oral communication 1 August 1997.

Schubert, P. and Heer, B. (1997) *Umweltverträgliche Verwertung von Mauerwerk-Baureststoffen.* Forschungsbericht Nr. F. 497. Institut für Bauforschung: Aachen.

Tebbe, H. and Hoffmann, S. (1995) *Möglicher Einsatz von Recycling-Bims in der Leichtbeton-Herstellung zur Ressourcenschonung und Primäreinergieeinsparung. Konsequenzen auf Druckfestigkeit, Rohdichte, Wärme- und Schallschutz und Auswirkungen auf die Ökobilanz (Energiebilanz) von Leichtbeton-Mauersteinen.* Forschungsbericht Nr. 9537. Materialprüfungs- und Versuchsanstalt: Neuwied.

VDI, Association of German Engineers (Hrsg.) (1993) *Guideline VDI 2243 – Design methods for reuse and recycling of technical products – Blatt 1,* VDI: Düsseldorf.

Weber, H. and Hullmann, H. (1991,1995) *PORENBETON HANDBUCH: Das Porenbeton Handbuch: Planen und Bauen mit System.* 1. ; 2. Aufl. Bauverlag: Wiesbaden.

Wesche, K. and Schulz, R.-R. (1982) 'Beton aus aufbereitetem Altbeton: Technologie und Eigenschaften', *Beton* 32(2), 64–68, 32(3), 108–112.

Willkomm, W. (1990) *Recyclinggerechtes Konstruieren im Hochbau: Recycling-Baustoffe einsetzen, Weiterverwertung einplanen.* Verlag TÜV Rheinland: Köln,

Walraven, J. (1988) 'Verbindungen im Betonfertigteilbau unter Berücksichtigung "stahlbaumäßiger" Ausführung', *Fertigteilbauforum Beilage in Betonwerk und Fertigteil-Technik* (1988), 20, 1–6.

5 Integrated life cycle design of materials

Toshio Fukushima

5.1 Design concept

Building materials are fundamental in providing shelter for people from the external environment and are heavily and widely used in construction. They may be organic, inorganic, metallic, or composite materials, and they may be traditional or new materials, but they must satisfy required performance levels for the components and elements comprising buildings in any given environment. To date, development of new building materials has been from the viewpoint of providing lightness, high strength and high durability in the case of structural materials such as steel, concrete and wood, and amenity characteristics of high functionality and flexible design in the case of such non-structural materials such as finishing polymer, glass and gypsum. As the importance of the global environmental problems threatening human existence in the coming twenty-first century becomes daily more appreciated, so environmental friendliness, as well as high levels of performance and functionality, is increasingly becoming as necessary a characteristic of building materials and components. Many building materials need to become environment-conscious materials (eco-materials) (Science and Technology Research and Development Bureau of Japan 1993, Yamamoto 1994, Nagai 1996). It is because carbon dioxide generation and consumption of energy when building materials are produced and the mass of various building materials wastes at the time of demolition, to no small extent cause burdens on the environment. Consequently the establishment of integrated life cycle design (which considers the balance between durability design for the assurance of long service life of building materials and/or components), and environment-conscious materials design – eco-materials design – (which considers, in advance, eco-balance performance such as recyclability and characteristics of the reduction of environmental load) is the important theme to be urgently undertaken (Sarja 1998). With durability design for the assurance of long service life of building materials and/or components in the total life cycle we can expect to reduce the amount and rate of generation of waste materials, and consequently make a reduction in the burdens on the global environment. However, we should also never fail to consider eco-materials design to assure the environmental-friendliness (eco-balance performance) of building materials and/or components in the total life cycle.

From another viewpoint, however, global environmental problems (such as the increase in the concentration of atmospheric CO_2 due to the massive consumption of fossil fuels, the generation of acid rain due to the air contamination by SO_x and NO_x gases, and the increase of dangerous shorter wavelength ultraviolet light due to the depletion of ozone layer caused by fluoro- and chloro-carbon gases) may accelerate the deterioration of building materials and components, and threaten to damage the long service life of houses and buildings as social capital, and massive building materials wastes are generated as the results of the unexpectedly fast deterioration. Figure 5.1 shows the triangular relationship between building materials, the global environment and the dwelling environment. Figure 5.2 shows also the triangular relationship among resource problems, deterioration problems and global environmental problems. We should, in future, use integrated life cycle design of building materials, taking into consideration the prediction of service life (PSL), life cycle cost analysis (LCC) and evaluation of the environmental load (life cycle assessment) (LCA) for the whole life cycle, including production, in-service use, disposal, and recycling and reuse. The importance of this type of design was emphasised in the RILEM Workshop on Environment and Building Materials held on Finland in 1995 (VTT 1995). Yoshio Kasai, Professor Emeritus at Nihon University proposed such an integrated life cycle design for buildings, including the optimum design of service life, maintenance, and recycling and reuse (Figures 5.3 and 5.4).

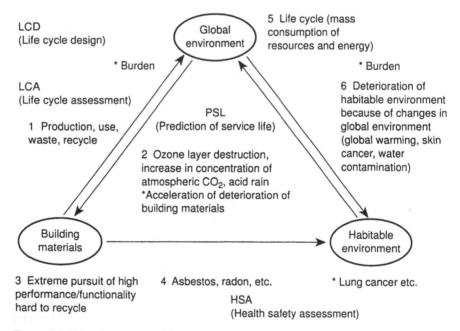

Figure 5.1 Triangluar relationship between building materials, the global environment and the dwelling environment

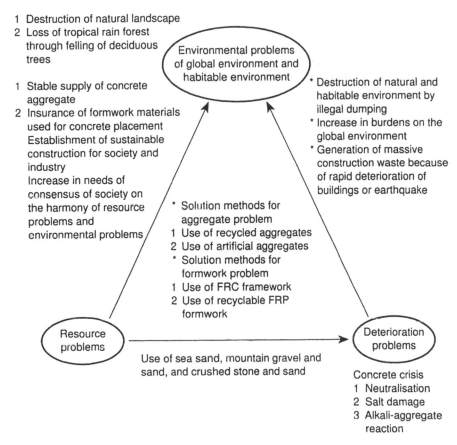

1 Destruction of natural landscape
2 Loss of tropical rain forest through felling of deciduous trees

1 Stable supply of concrete aggregate
2 Insurance of formwork materials used for concrete placement
Establishment of sustainable construction for society and industry
Increase in needs of consensus of society on the harmony of resource problems and environmental problems

Environmental problems of global environment and habitable environment

* Destruction of natural and habitable environment by illegal dumping
* Increase in burdens on the global environment
* Generation of massive construction waste because of rapid deterioration of buildings or earthquake

* Solution methods for aggregate problem
1 Use of recycled aggregates
2 Use of artificial aggregates
* Solution methods for formwork problem
1 Use of FRC framework
2 Use of recyclable FRP formwork

Resource problems

Deterioration problems

Use of sea sand, mountain gravel and sand, and crushed stone and sand

Concrete crisis
1 Neutralisation
2 Salt damage
3 Alkali-aggregate reaction

Figure 5.2 Triangular relationship between resource problems, deterioration problems and environmental problems

5.2 Environmental design procedure

How to develop sustainable eco-cities and eco-societies taking into consideration environmental friendliness is becoming an urgent problem in the twenty-first century. In order to realise a sustainable construction industry whose activities have a great influence on the global environmental load, we have to consider the harmony among the three aspects of the environment, the economy and the long service life/safety. In the field of construction, the development of environmentally-conscious buildings (eco-buildings) and environmentally-conscious building component technology (eco-technology) is the basis of sustainability (Kasai 1995, Fuksuhima 1997b).

From the viewpoint of building materials, we should take into account the balance between the development of high performance and high function of building materials and/or components, structural systems using new advanced materials (such as fibre reinforced plastics (FRP) and fibre reinforced cement composites (FRC)) and eco-materials design. Consequently, the fundamental drive

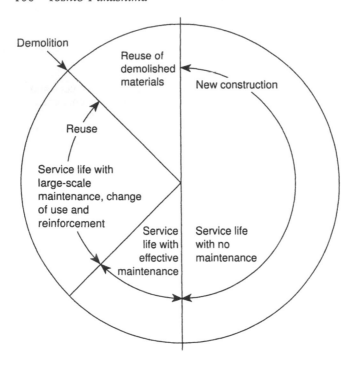

Figure 5.3 Life, maintenance and recycling of buildings

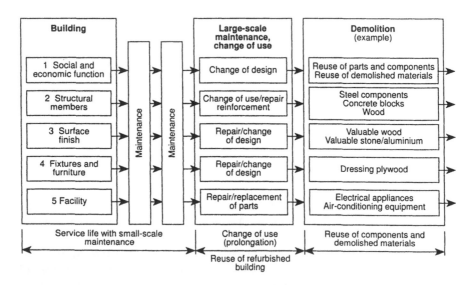

Figure 5.4 Service life and its prolongation of buildings and reuse of demolished components

of environmental design procedure is how to achieve environmentally-conscious life cycle design (including eco-materials design) – eco-life cycle design – to develop sustainable eco-cities and eco-buildings, by focusing on composite materials. This is because, from a fundamental viewpoint, the ways of using materials in architecture and civil engineering assumes the use of composite materials or component, and the use of a simple non-composite materials is very rare.

Existing composite structural building materials and components such as steel reinforced concrete (RC) and continuous fibre reinforced plastic reinforced concrete (FRPRC) have many excellent characteristics. An extreme concentration on high performance characteristics such as lightness, high strength and high corrosion resistance, however, becomes an obstacle to the recycling of these structural composite materials and/or components when demolishing after designed service life. We should subsitute these structural composite materials and/or components with new recycling-oriented ones, taking into account the harmony of long service life with recyclability, and using eco-materials technology. FRP reinforced lightweight precast concrete is taken as an example of applying eco-materials design, because it uses both FRP and concrete, each of which are currently considered to be most difficult to recycle. Some fundamental eco-materials technology to establish this type of eco-materials design should be pursued. The fundamental concepts of eco-materials design of building structural composite materials and components are shown in Figures 5.5–5.8 (Fukushima, Yanagi and Maeda 1995, Fukushima, Kojima, Yanagi and Yoshiizaki 1995, Fukushima *et al.* 1998). How to establish the harmony of long service life with recyclability is an important point of this eco-materials design. Figure 5.5 is an outline of eco-materials design. Figure 5.6 is an outline of the environmentally-conscious life cycle design (eco-life cycle design) of building structural composite materials, taking into consideration the harmony of long service life with recyclability. Figure 5.7 is the design of selection and evaluation of FRPRC for effective use as building materials in given building elements. Figures 5.8–5.10 are the concepts of basic technology for realising eco-materials design.

5.3 Evaluation indicators of environmental-friendliness (eco-balance performance) (Fukushima and Shiire 1998)

Outline

Recently, the establishment of sustainable society taking into account the balance between the environment, the economy and safety/long service life has been becoming an urgent problem for the twenty-first century. As construction and the building environment have been recognised as the biggest consumers of natural resources and energy, so all the more efforts have to be made to accomplish sustainable construction. Meanwhile, conversion of building materials into eco-materials is needed because the environmental burdens imposed by the massive use in various ways of buildings materials are very serious throughout their life cycle.

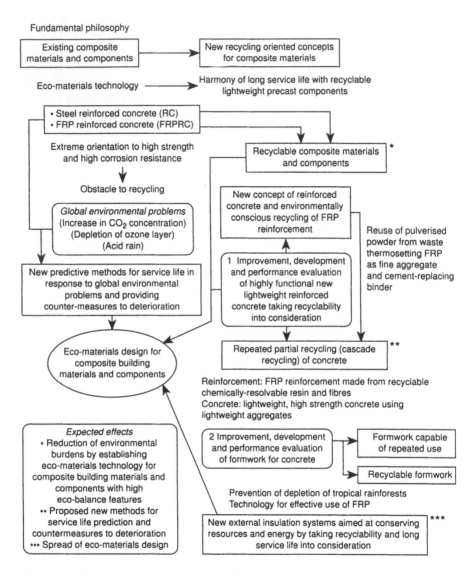

Figure 5.5 Fundamental concepts for environmentally-concious materials design (eco-materials design for composite structural building materials and components)

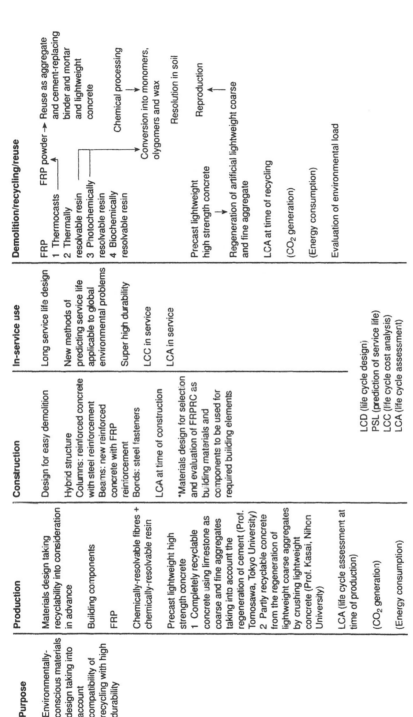

Purpose	Production	Construction	In-service use	Demolition/recycling/reuse
Environmentally-conscious materials design taking into account compatibility of recycling with high durability	Materials design taking recyclability into consideration in advance	Design for easy demolition	Long service life design	FRP FRP powder → Reuse as aggregate and cement-replacing binder and mortar and lightweight concrete
	Building components	Hybrid structure Columns: reinforced concrete with steel reinforcement Beams: new reinforced concrete with FRP reinforcement Bonds: steel fasteners	New methods of predicting service life applicable to global environmental problems	1 Thermocasts 2 Thermally resolvable resin 3 Photochemically resolvable resin 4 Biochemically resolvable resin — Conversion into monomers, olygomers and wax — Chemical processing — Resolution in soil
	FRP	LCA at time of construction	Super high durability	
	Chemically-resolvable fibres + chemically-resolvable resin	*Materials design for selection and evaluation of FRPRC as building materials and components to be used for required building elements	LCC in service	Precast lightweight high strength concrete
	Precast lightweight high strength concrete 1 Completely recyclable concrete using limestone as coarse and fine aggregates taking into account the regeneration of cement (Prof. Tomosawa, Tokyo University) 2 Partly recyclable concrete from the regeneration of lightweight coarse aggregates by crushing lightweight concrete (Prof. Kasai, Nihon University)		LCA in service	Regeneration of artificial lightweight coarse and fine aggregate — Reproduction
	LCA (life cycle assessment at time of production)			LCA at time of recycling
	(CO_2 generation)			(CO_2 generation)
	(Energy consumption)			(Energy consumption)
				Evaluation of environmental load

LCD (life cycle design)
PSL (prediction of service life)
LCC (life cycle cost analysis)
LCA (life cycle assessment)

Figure 5.6 Environmentally-conscious materials design (eco-material design) of pre-cast lightweight concrete with continuous fibre reinforced plastic (FRP) reinforcement

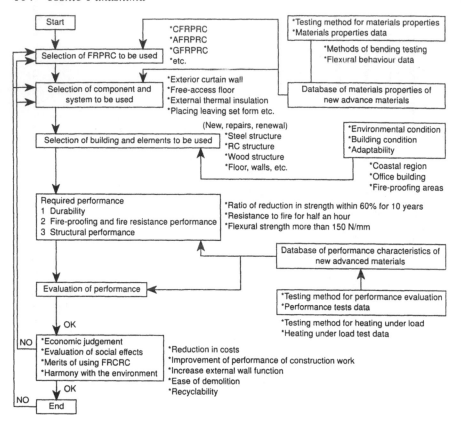

Figure 5.7 Selection and evaluation system for FRPRC for effective use in construction

Steel reinforced concrete (RC) and continuous fibre reinforced concrete (FRPRC) are the two main composite structural building components, and despite their long service life and high performance in service, they are difficult to demolish and recycle after the designed service life, and are liable, to a great extent, to impose burdens on the environment. Consequently we have to develop eco-materials design and eco-life cycle design (and develop the technology to realise their design) for these composite structural building components throughout their life cycle of production, construction, in-service use, demolition, disposal/recycling/reuse. In order to evaluate eco-balance performance in the total life cycle of these structural composite materials, we should establish evaluation indicators of environmental harmony.

Because of this, based upon the new consideration from the viewpoint of the life cycle assessment (LCA) concept, we aim at developing eco-life cycle design which takes account of the balance of various types of performance demanded from building structures and components. By establishing evaluation indicators of environmental harmony (eco-balance performance) – such as life cycle inventory (LCI), life cycle environmental impact analysis (LCIA) recyclability indicator

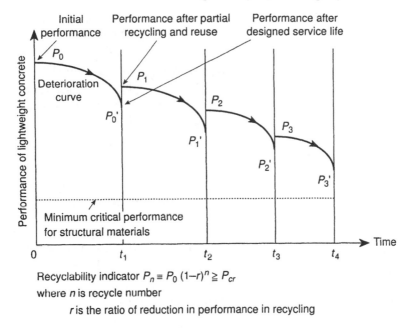

Recyclability indicator $P_n \equiv P_0 (1-r)^n \geq P_{cr}$

where n is recycle number

r is the ratio of reduction in performance in recycling

Figure 5.8 Repeated partial (cascade) recycling of lightweight precast concrete

(RI), waste unoccupancy indicator (WUI) – giving a basis for the establishment of sustainable eco-buildings and eco-cities in the twenty-first century.

Life cycle design of composite structural building components

In order to establish rational methods of using composite structural building components to acheive compatibility of long service life with eco-balance performance, we have to achieve life cycle design from the viewpoint of life cycle assessment, for the total life cycle process of production, construction, in-service use, demolition, disposal/recycling/reuse. This is based on three main methods: prediction of service life (PSL), life cycle cost analysis (LCC), and life cycle environmental impact assessment (LCIA) (Fukushima 1997). Figures 5.11 and 5.12 show the fundamental schemes of such life cycle design.

Life cycle assessment (LCA) of composite structural building components

In order to rationally evaluate eco-balance performance as a basis of life cycle design for the total life cycle process of composite structural building components, we have to establish comprehensive evaluation methods based on life cycle inventory (LCI), global_life cycle environmental impact analysis (LCIA),

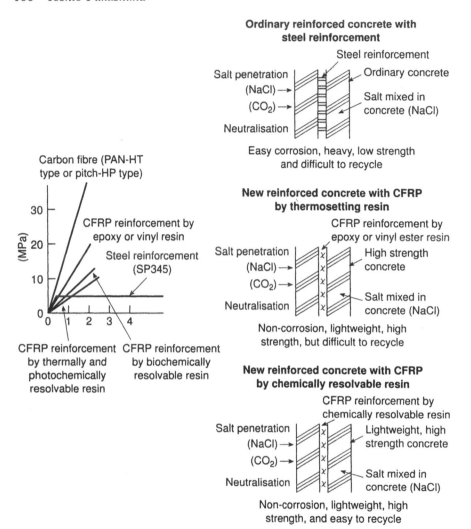

Figure 5.9 Comparison of new reinforced concrete with FRP reinforcement with traditional reinforced concrete with steel reinforcement

recyclability indicator (RI), waste unoccupancy indicator (WUI). These four evaluation indicators will be examined in turn.

Life cycle inventory (LCI)

In order to discuss the eco-balance performance of composite structural building components such as steel reinforced concrete (RC) and continuous fibre reinforced concrete (FRPRC), we need life cycle design (LCD) of the total life cycle process

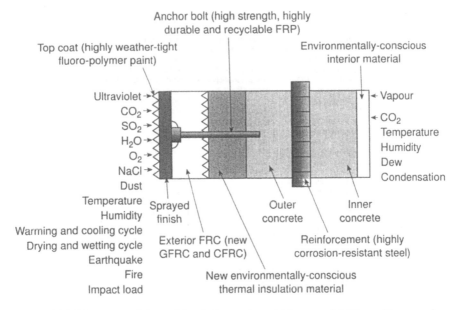

Figure 5.10 New external thermal insulation systems taking recyclability and long service life into consideration

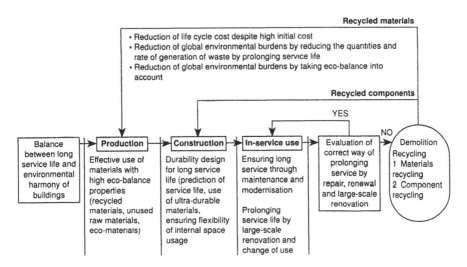

Figure 5.11 Life cycle design aimed at compatibility of long building service life with environmental harmony (eco-balance performance)

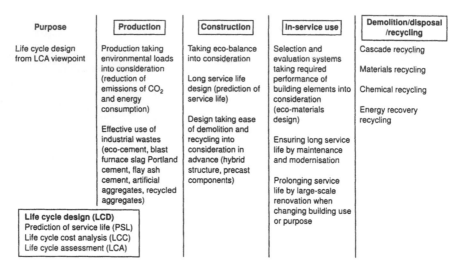

Purpose	Production	Construction	In-service use	Demolition/disposal /recycling
Life cycle design from LCA viewpoint	Production taking environmental loads into consideration (reduction of emissions of CO_2 and energy consumption) Effective use of industrial wastes (eco-cement, blast furnace slag Portland cement, flay ash cement, artificial aggregates, recycled aggregates)	Taking eco-balance into consideration Long service life design (prediction of service life) Design taking ease of demolition and recycling into consideration in advance (hybrid structure, precast components)	Selection and evaluation systems taking required performance of building elements into consideration (eco-materials design) Ensuring long service life by maintenance and modernisation Prolonging service life by large-scale renovation when changing building use or purpose	Cascade recycling Materials recycling Chemical recycling Energy recovery recycling
Life cycle design (LCD) Prediction of service life (PSL) Life cycle cost analysis (LCC) Life cycle assessment (LCA)				

Figure 5.12 Procedures for environmentally-conscious life cycle design for composite structural building materials and components

of production, in-service use, demolition, disposal/recycling/reuse, based upon three procedures of PSL, LCC and LCA. To begin with, we summarise the concept of life cycle inventory (LCI) as the starting point of LCA. For concrete, steel, fibre reinforced thermoset (FRTS), fibre reinforced thermoplastic (FRTP), the life cycle CO_2 emission amount ($LCCO_2$) should be calculated. Further using these values $LCCO_2$(concrete), $LCCO_2$(S), $LCCO_2$ (FRTS) and $LCCO_2$ (FRTP), the $LCCO_2$ should be calculated for RC and FRTSRC and FRTPRC. By comparing these values of $LCCO_2$(RC), $LCCO_2$ (FRTSRC), $LCCO_2$ (FRTPRC), we are able to discuss the eco-balance performance of these structural composite components.

$LCCO_2$ (concrete) = L production + L construction + L in-service + L demolition + L recycle

$LCCO_2$ (FRTS) = L production + ... + L recycle ~ 0.66 kg / year

$LCOO_2$ (FRTP) = L production + ... + L recycle ~ 0.47 kg / year

$LCCO_2$ (S) =-L production + ... + L recycles

$LCCO_2$ (FRTSRC) = $LCCO_2$ (concrete) + $LCCO_2$ (FRTS) + $LCCO_2$ (correlation function)

$LCOO_2$ (FRTPRC) = $LCCO_2$ (concrete) + $LCCO_2$ (FRTP) + $LCCO_2$ (correlation function)

$LCCO_2$ (RC) = $LCCO_2$ (concrete) + $LCCO_2$ (S) + $LCCO_2$ (correlation function)

*Comparison: $LCCO_2$ (FRTSRC) > = < $LCCO_2$ (RC)
      $LCOO_2$ (FRTPRC) > = < $LCCO_2$ (RC)
      $LCCO_2$ (FRTPRC) > = < $LCCO_2$ (FRTSRC)

Life cycle environmental impact analysis (LCIA)

In order to try to assure the practicability of LCA, we should never fail to consider the life cycle environmental impact analysis (LCIA) for composite structural building components, as currenlty many reports of LCA contain only LCI information and without LCIA data LCA is not complete in any real sense. For example, we should consider LCIA of composite structural building components such as RC and FRPRC based upon $LCCO_2$ and whether the concentration of the global atmospheric CO_2 increases or decreases by using these building structural components, as shown in Figure 5.13.

Recyclability indicator (RI)

Many people stress the importance of recycling as well as the preservation of natural resources and the conservation of energy in order to respond effectively to global environmental problems. But, so far, quantitative evaluation indicators of recyclability are not clear. We would like to set out evaluation indicators for repeated partial recycling of FRPRC using pulverised waste FRP powder as cement-replacing binder and fine aggregate (cascade recycling) as shown in Figure 5.14.

The main point of this cascade recycling is to partially recycle concrete several times until critical performance is reached, taking into account both rate of reduction of performance and recycling number.

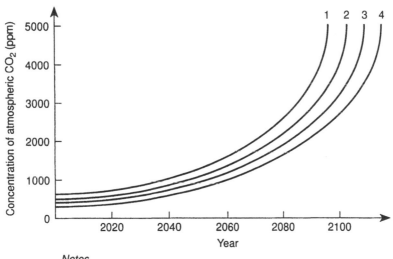

Notes
1 20% increase in annual CO_2 emission until 2000
2 No effective measures to reduce annual CO_2 emission until 2000
3 20% reduction in annual CO_2 emission until 2000
4 40% increase in annual CO_2 emission until 2000

Figure 5.13 Prediction of changes in atmospheric CO_2 in the twenty-first century

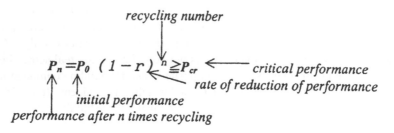

recycling number

$$P_n = P_0 \ (1-r)^n \geq P_{cr} \longleftarrow critical\ performance$$
$$\longleftarrow rate\ of\ reduction\ of\ performance$$

initial performance

performance after n times recycling

Figure 5.14 Evaluation indicators for repeated partial recycling of FRPRC

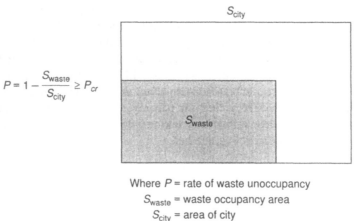

S_{city}

$$P = 1 - \frac{S_{waste}}{S_{city}} \geq P_{cr}$$

S_{waste}

Where P = rate of waste unoccupancy
S_{waste} = waste occupancy area
S_{city} = area of city
P_{cr} = critical rate of waste unoccupancy

Figure 5.15 Relationship of city area and waste occupancy area

$$P_3\ (\text{concrete}) = 100\ MP_a(1-0.20)^3 = 100 \times 0.512 = 51.2\ MP_a > = 30MP_a$$
$$P_4\ (\text{FRTP}) = 250\ MP_a\ (1-0.10)^4 = 250 \times 0.6561 = 164\ MP_a > = 100\ MP_a$$

According to this approach, concrete and FRTP can be recycled three and four times, respectively.

Waste unoccupancy indicator (WUI)

When waste occupies a considerable area of the habitable environment of a city, we think that environmental burdens have become significant. Consequently, we define the waste unoccupancy indicator (WUI) as shown in Figure 5.15.

According to this indicator, environmental burdens become very serious if the rate of waste unoccupancy P becomes below the critical value P_{cr} (for example 60%).

5.4 Definition and optimisation of service life

The service life of building materials used in buildings should be as long as possible during in-service use to conserve resources and energy, so as to reduce burdens on the environment. However, compatibility of the service lives of different components should be considered: structural materials and components should have long durability during the designed service life, but non-structural or finishing materials, such as gypsum and polymer, will be renewed or replaced at intervals in order to maintain the building as a whole in good condition.

In many cases, the service life of reinforced concrete (RC) buildings in ordinary atmospheric surroundings in mild climates is determined by the progress of carbonation of the concrete and corrosion of the reinforcing steel (Sciessl 1976, Architectural Institute of Japan 1982) except in special cases such as deterioration due to salt damage in severe maritime environments and frost damage in very cold climates.

At present there are very few concrete buildings using continuous fibre reinforced plastic reinforced concrete (FRPRC), except for special cases where high corrosion resistance is required of reinforcements – such as the Japanese Showa Base in the Antarctic. In these cases, the deterioration of FRP reinforcement becomes the basis of physical service life of FRPRC buildings.

Figure 5.16 shows the comparison of methods of defining physical service life of off-form type reinforced concrete (RC) and continuous fibre reinforced plastic reinforced concrete (FRPRC) composite structural components.

In RC, we can reasonably define physical service life as t_2, when the occurrence of cracks and spalling of the cover concrete creates a safety hazard. This is because time t_1, when the neutralisation of the cover concrete reaches the reinforcing steel, is set too much on the safety side, and time t_3, when structural safety of RC components becomes severely damaged, is set too much on the dangerous side. If we make effective use of surface-finishing materials, however, the service life of reinforced concrete buildings can be prolonged by the suppressive effects of surface finishing materials against the progress of carbonation of the concrete and corrosion of the reinforcing steel Fukushima and Fukushi 1992, 1993, 1997, Fukushima, Tomosawa, Fukushi and Tanaka 1993, 1999a, 1999b, Fukushima 1997b). It is assumed that surface finishing materials have high durability for a long time, and do not deteriorate as much as in the case of fluoropolymer, or that even if they do deteriorate, the periodic replacement or renewal will be carried out.

For FRPRC, however, we should establish new rational methods of predicting service life. We would like to tentatively propose defining the physical service life of FRPRC as t_1 when neutralisation of the cover concrete reaches the FRP reinforcement as the result of the progress of carbonation, because, severe deterioration of the FRP reinforcement then rapidly progresses due to bacteria attack.

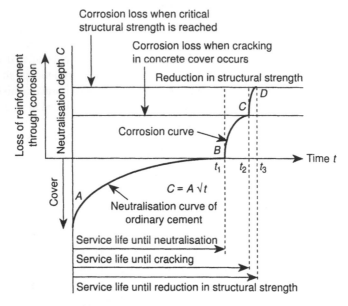

a) Ordinary reinforced concrete
(steel reinforcement + ordinary concrete)
(service life best determined by t_2)

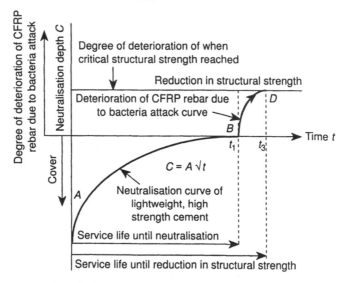

b) New reinforced concrete
(CFRP reinforcement made from biochemically resolvable resin + lightweight, high strength concrete
(service life best determined by t_1)

Figure 5.16 Comparison of predicted service life between new reinforced concrete with FRPRC reinforcement and ordinary reinforced concrete with steel reinforcement

5.5 Design for a defined service life

We can define the designed service life, based upon the proposed use of buildings, for example, at 30 years, 50 years and 100 years. To ensure the designed service life, the appropriate selection of building materials, good construction practices and periodic maintenance and renewal are needed. Figure 5.17 shows the design of a defined service life of reinforced concrete buildings by effective use of the suppressive effects of surface finishing materials and their periodic maintenance and renewal and replacement.

5.6 Examples

The following four examples show elementary techniques to realise these environmental design methods.

1   Reuse method of waste thermosetting FRP (FRTS) rebar by pulverising as fine aggregate and cement-replacing binder in repeated partial recycling (cascade recycling) of concrete (Fukushima 1998, Fukushima, Yanagi and Maeda 1995, Yanagi, Kojima and Fukushima 1996).

   Table 5.1 shows the experimental results of concrete mixed with pulverised waste FRP powder (glass short-cut fibre reinforced unsaturated polyester), and Table 5.2. shows the experimental results of fluidised concrete mixed with pulverised waste FRP powders. By tests of accelerated carbonation and freezing-thawing, we confirmed good durability of these recycled aggregate concretes. It was found from these experimental results that concrete mixed with pulverised waste FRP powders show material characteristics comparable to plain concrete. However, we should establish effective techniques to pulverise waste thermosetting FRP rebar into powders after designed service life of FRPRC.

2   Application of recyclable thermoplastic FRP for reinforcements for precast concrete structures (Fukushima 1997, Fuksuhima *et al.* 1998, Fukushima, Sakayama and Hashimoto 1997a, 1997b).

   In order to convert FRPRC into eco-materials, recyclable non-metallic reinforcements for concrete structures, made of continuous carbon and glass fibre reinforced thermoplastics (CFRTP and GFRTP) have recently been developed (Photo 5.1).

   Table 5.3 shows the materials properties and test methods of these newly-developed recyclable continuous carbon and glass fibre reinforced thermoplastics (FRTP).

   We confirmed that these CFRTP and GFRTP can be recycled three times by recyclability testing. The dynamic behaviours of these FRTP were examined by repeated tensile loading testing. It was found that these newly-developed FRTP are applicable for non-metallic reinforcements, having the characteristics of environment friendliness as well as satisfactory dynamic characteristics necessary for precast concrete reinforcements.

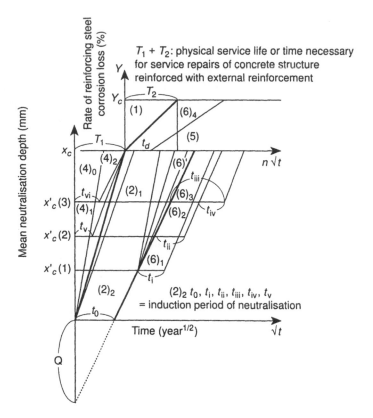

Progress of corrosion
$Y^n = a(t - t_d) : n \geq 1$
when Y_c = critical rate corrosion loss
(when cracking and spalling of cover
concrete occurs because of pressure
from the expansion of corrosion rust)
t_d = induction period of corrosion

Progress of neutralisation
$0 \geq t > t_0 : X = 0$
$t_0 \geq t \, X = K_e \sqrt{t} - Q$

where X_c is the critical neutralisation
depth (when neutralisation of cover
concrete reaches the surface of the
reinforcing steel)
$X'_c \approx 1/3_c, 1/2_c, 2/3_c$
Q = depth of delay of neutralisation

(1) Off-form concrete
(2) Suppressive effects of surface finishing
     work from initial stage
$(2)_1$ Suppressive effects or thickness of
     surface finishing are large
$(2)_2$ Suppressive effects or thickness of
     surface finishing is small
(3) Suppressive effects of surface finish
     decrease because of deterioration
(4) Progress of neutralisation is suppressed
     up to the service life of off-form concrete,
     but it progresses unexpectedly quickly
(5) Progress of reinforcing steel corrosion in
     neutralised concrete is suppressed by
     finishing materials
(6) Repair and renewal of surface finishing
     materials carried out at appropriate
     intervals

Figure 5.17 Setting the physical service life of reinforced concrete components by taking
account of the suppressive effects of surface finishing materials

Table 5.1 Experimental results of recycled concrete mixed with pulverised FRP powder (curing time 28 days)

| FRP ratio | Slump | Water cement ratio | Sand aggregate ratio | Unit water weight | Unit weight | | | | Air content | Compressive strength | Young modulus |
| | | (S/A) | | | Cement | Sand | FRP | Gravel | | f_c | E |
(%)	w/c (cm)	(%)	(%)	(g/2L) (kg/m³)					(%)	(N/mm²)	(×10³n/mm²)
0	19.0	60	45.0	175	292	685	–	896	4.8	22.1	16.1
1	19.9	60	45.0	175	292	678	7	896	8.6	14.2	12.8
5	18.5	60	45.0	175	292	651	34	896	9.3	14.0	12.1

Table 5.2 Experimental results of fluidised concrete mixed with pulverised FRP powder (curing time 7 days)

| Types of FRP | FRP ratio | Flow | W/P* | Unit water weight | Unit weight | | | | High performance water-reducing reagent ratio | Air content | Compressive strength |
| | | | | | Cement | Sand | FRP | Gravel | | | f_c |
	(%)	(mm)	(%)	(kg/m³)	(g/2L)				(%)	(%)	(N/mm²)
–	0	495	35.0	51.8	486	860	-	806	2.0	3.6	64.7
Pulverised powder	5	405	35.0	51.8	462	860	24	806	2.0	2.9	63.0
Fly ash	5	418	35.0	51.8	462	860	24	806	2.0	3.9	63.2

Table 5.3 Material properties and test methods for thermoplastic FRPs

Test items	Test method	Unit	PVC/GF	PC/GF	PVC/CF	PC/CF
Specific gravity	ASTM D792	–	1.50	1.34	1.51	1.35
Tensile strength	ASTM D638	N/mm^2 (Kgf/cm^3)	129 (1320)	118 (1200)	347 (3450)	347 (3540)
Flexural strength	ASTM D790	N/mm^2 (Kgf/cm^3)	173 (1770)	169 (1720)	189 (1930)	312 (3180)
Flexural elasticity	ASTM D790	N/mm^2 (Kgf/cm^3)	7771 (79300)	3125 (32500)	21000 (212000)	21000 (212000)
Izod type impact strength	ASTM D256	J/m (Kgf/cm^3)	304 (31)	774 (79)	304 (31)	304 (31)

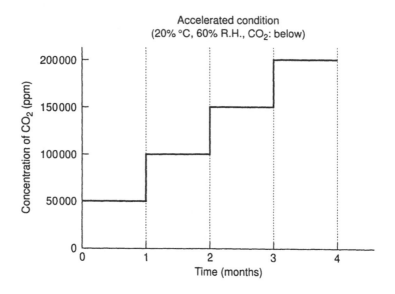

Figure 5.18 Method of accelerated carbonation testing by step-response modelling the tendency to increase the concentration of atmospheric CO_2

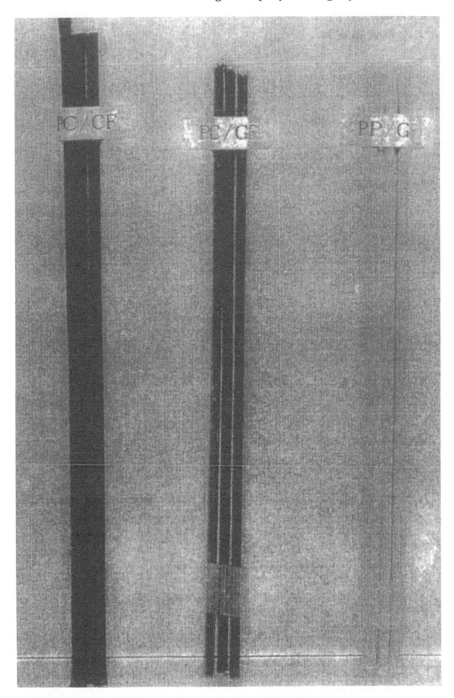

Photo 5.1   Outline of newly developed recyclable continuous fibre reinforced thermoplastics

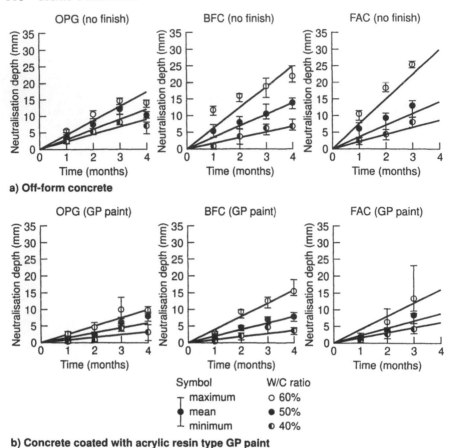

Figure 5.19   Progress of neutralisation of various types of concrete by step-response method

3   Accelerated carbonation testing of concrete by step-response methods as a
new method of prediction of service life of RC and FRPRC corresponding to
the global environmental problems (Fukushima, Yoshizaki and Hayashi 1995,
1996, 1997).

Accelerated carbonation testing was performed based on step-response
methods by setting the environmental CO_2 concentration as shown in Figure
5.18 modelling the tendency to increase the concentration of the atmospheric
CO_2 for various types of concrete (ordinary Portland cement concrete (OPC),
Portland blast-furnace slag cement concrete (BFC), and Portland fly ash
cement concrete (FAC) with and without acryl resin-type polymeric surface
finish. Linear rather than parabolic law, concerning time was observed for
the first time by this new accelerated carbonation test method as shown Figure
5.19. The theoretical explanation considering saturation term for the influence
of CO_2 concentration on the progress of neutralisaton fitted this linear law
well.

4 Prediction and evaluation methods of new external thermal insulation systems taking into consideration the compatibility of recyclability with long service life and the preservation of energy and the conservation of resources (VTT 1995, Fukushima, Yanagi, Uezono and Mothashi 1994, Fukushima, Shiire, Yanagi and Kurauchi 1999).

New external thermal insulation systems using PAN (polyamide-nitrate)-type carbon short-cut fibre reinforced cement composites (PAN-CFRC) as exterior materials were well evaluated by accelerated durability testing based upon repeated cycles of warming and cooling with water spray.

A trial to convert new external thermal insulation systems using pitch-CFRC as exterior materials (CFRC) into environment-conscious ones, dynamic behaviours of pitch-CFRC using eco-cement and pulverised waste FRP powders as replacement materials for ordinary Portland cement, and also using recycled fine aggregate as replacement materials for river sand were examined and it was found that they have dynamic behaviours comparable to ordinary CFRC using only virgin resource materials. The results obtained a main root for conversion of exterior CFRC into eco-materials and new external thermal insulation systems into environment-conscious ones.

References

Architectural Institute of Japan (1982) Japanese Architectural Standard Specification for Reinforced Concrete Work JASS 5, Architectural Institute of Japan: Tokyo.

Fukushima, T. (1997a) 'Theoretical analysis of suppressive effects of polymeric surface finishing materials on corrosion of reinforcing steel in neutralized concrete', in Ohama, Y. and Puterman, M. (eds) *Adhesion between Polymers and Concrete*, Proc. 2nd Interntl. RILEM Sym. (ISAP '99), 14–17 September, Dresden. RILEM: Paris.

Fukushima, T. (1997b) 'Environment-conscious materials design (ecomaterials design) of building structural composite materials and components and/or systems – as a basis of establishment of sustainable eco-cities and eco-buildings' *Proc. 3rd Interntl. Conf. Ecomaterials*, 10–12 September, Tsukuba. Ecomaterial Forum: Tokyo.

Fukushima, T. (1998) 'Life cycle design and ecomaterials technology of building structural composite components for sustainable construction', *Proc. of Symposium A: Materials and Technologies for Sustainable Construction, CIB World Building Congress*. CIB/RILEM: Gävle.

Fukushima, T. and Fukushi, I. (1992) 'Quantitative evaluation of the influence of polymeric surface finishing materials on the progress of neutralization of concrete: prediction of service life of external vertical walls of reinforced concrete buildings (Part 2)', *Journal of Structural Engineering of Architectural Institute of Japan* 434, 1–12 (in Japanese).

Fukushima, T. and Fukushi, I. (1993) 'Durability design of reinforced concrete buildings: quantitative evaluation of suppressive effects of surface finishing materials on the progress of neutralization (coronation) of concrete considering their deterioration and methods of setting equivalent thickness of concrete cover' *CementConcrete* 553, 50–60 (in Japanese).

Fukushima, T. and Fukushi, I. (1997) 'Quantitative evaluation of suppressive effects of polymeric surface coating materials on carbonation of concrete: study and prediction on service life of reinforced concrete buildings (Part 2)', in Ohama. Y., Kawakami, M. and Fukizawa, K. (eds), *Polymers in Concrete*. Proc. 2nd East Asian Sym. Polymers in Concrete, 11–13 May, Koriyama. E&FN Spon: London.

Fukushima, T., Kojima, A., Yanagi, K., Yoshizaki, Y. *et al.* (1995) 'Ecomaterials design of building structural composite materials and/or components considering recyclability and its elemental technology', *Proc. Interntl. Ecomaterials Conf.* 10–15 September, Xian.

Fukushima, T., Sakayama, K. and Hashimoto, S. (1997a) 'Dynamic behaviors of continuous carbon and glass fiber reinforced thermoplastics (CFRP and GFRP) for recyclable non-metallic reinforcement for concrete structures', *Proc. 3rd Interntl. Sym. Non-metallic Reinforcement for Concrete Structures*, 14–16 October, Sapporo. Japan Concrete Institute: Tokyo.

Fukushima, T., Sakayama, K. and Hashimoto, S. (1997b) 'Application of recyclable thermoplastic FRP for reinforcement for lightweight precast concrete', in Miyano, Y. and Yamabe, M. (eds) *Challenging to Advanced Materials and Processing Technology Aiming for New Industrial Applications*. Proc. 5th Japan Interntl. SAMPE Sym., 28–31 October, Tokyo, Japan. Japan Chapter of SAMPE: Yokohama.

Fukushima, T. and Shiire, T. (1998) 'Evaluation indicators of environmental harmony of building structural composite components', *Proc. 3rd Interntl. Conf. Ecobalance*, 25–27 November, Tsukuba. Ecomaterial Forum: Tokyo.

Fukushima, T., Shiire, T., Yanagi, K., Sone, T. and Kurauchi, H. (1999) 'Ecomaterials design of new external thermal insulation systems and performance evaluation: Part 3. Conversion methods of exterior short-cut fiber reinforced cement composites into ecomaterials', *Proc. Annual Meeting of Architectural Institute of Japan, A-1 (Materials and Construction Works)* (in Japanese). Architectural Institute of Japan: Tokyo.

Fukushima, T., Tomosawa, F., Fukushi, I. and Tanaka, H. (1993) 'Quantitative evaluation of suppressive effects on corrosion of reinforcing steel in neutralized concrete', in Nagataki, S., Nireki, T. and Tomosawa, F. (eds), *Durability of Building Materials 6*, Proc. 6th Interntl. Conf., 26–31 October, Omiya. E&FN Spon: London.

Fukushima, T., Tomosawa, F., Fukushi, I. and Tanaka, H. (1999a) 'Experimental study on suppressive effects of surface finishing materials on corrosion of reinforcing steel in neutralized concrete', *Concrete Research and Technology*, 11(2), 83–94 (in Japanese).

Fukushima, T., Tomosawa, F., Fukushi, I. and Tanaka, H. (1999b) 'Theoretical study on suppressive effects of surface finishing materials on corrosion of reinforcing steel in neutralized concrete', *Concrete Research and Technology*, 11(2), 108 (in Japanese).

Fukushima, T., Yanagi, K. and Maeda, T. (1995) 'Environmental conscious materials design of lightweight precast concrete components with recyclable FRP rebars', in Ohama, Y. (ed.) *Disposal and Recycling of Organic and Polymeric Construction Materials*. Proc. Interntl. RILEM Workshop, 26–28 May, Tokyo. E&FN Spon: London.

Fukushima, T., Yanagi, K., Uezono, M. and Motohashi, K. (1994) 'New external thermal insulation systems using new carbon chopped fiber reinforced cement composites', *Proc. Interntl. Sym. Managing, Maintenance and Modernization of Building Facilities*, 26–28 October.

Fukushima, T., Yoshizaki, Y. and Hayashi, T. (1995) 'Accelerated carbonation test by step response method and theoretical explanation of the results', *JCA Proceedings Cement & Concrete* 49, 692–697 (in Japanese).

Fukushima, T., Yoshizaki, Y. and Hayashi, T. (1996) 'Accelerated carbonation test by step response method', in Sjorstrom, C. (ed.) *Durability of Building Materials & Components*

7, Proc. 7th Interntl. Conf. Durability of Building Materials and Components, 19–23 May, Stockholm. E&FN Spon: London.

Fukushima, T., Yoshizaki, Y. and Yakahashi, K. (1997) 'Theoretical analysis of the results of accelerated carbonation test by step response method', *JCA Proceedings Cement & Concrete* 51, 672–677 (in Japanese).

Fukushima, T., *et al.* (1998) 'Environment-conscious life cycle design of continuous fiber reinforced concrete and elemental materials technology', *Proc. Sym. Continuous Fiber Reinforced Concrete* (in Japanese), Japan Concrete Institute: Tokyo.

Kasai,Y. (1995) 'Trial view to life, maintenance and recycling of buildings', *Building Technology and Works*, 102–108 (in Japanese).

Nagai, K. (1996) *Recycling of Civil Engineering and Building Materials* (in Japanese). Ecomaterials Series. Chemical Industry Publishers: Tokyo.

Sarja, A. (1998) 'Integrated life cycle design of materials and structures', *Proc. of Symposium A: Materials and Technologies for Sustainable Construction, CIB World Building Congress*. CIB/RILEM: Gävle.

Science and Technology Research and Development Bureau of Japan (1993) *Report of Fundamental Research Based upon the Budget in 1992 Fiscal Year for Promotion and Adjustment of Science and Technology: Ecomaterials for the Preservation of the Global Environment* (in Japanese). Research Development Bureau of Science and Technology of Japan: Tokyo.

Sciessl, P. (1976) 'Zur Frage der Zulässingen Rissbreit und der Erfoldichen Betondeckung in Stahlbetonbau unter besonderer Berücksichting der Karobonatisierung des Betons', *Deutscher Ausschuss für Stahlbeton*, 255, 27–83.

VTT (1995) *Proceedings of RILEM Workshop on Building Materials and Environment*, 18–19 November, Otaniemi. VTT (Center for Building Technology): Espoo.

Yamamoto, R. (1994) 'Materials revolution gentle to the global environment', *Ecomaterials Review*, 207–215 (in Japanese).

Yanagi, K., Kojima, A. and Fukushima, T. (1996) 'Physical properties of concrete containing scrapped FRP fine powder', *JCA Proceedings of Cement & Concrete*, Paper 161 (in Japanese).

6 Conclusions and needs for further development

6.1 Conclusions

The incorporation of integrated design principles into practical design is a fairly extensive process, in which not only is the work of structural engineers changing, but the co-operation between structural engineers, architects, building services system designers and other partners of construction and use also has to be developed. This applies especially to co-operation with clients and architects which is important in order to effectively utilise the expertise of structural engineers in the most decisive, creative and conceptual phases of design. This kind of co-operation also helps the clients to realise the benefits of investing slightly more in the design. The first part of the design, which is performance-oriented, will be carried out by architects and technical designers in close co-operation both with each other and with the client and users. The second part, which is a concretising phase, is made in a close team of technical designers and manufacturers. The concretising phase is often connected to specific building concepts of contractors and their suppliers. In this way the current problem of diversified design and manufacturing processes can be avoided without compromising the functional and performance requirements and other requirements for the life cycle use of the building, which are discussed above.

Designers are increasingly participating in product development of prefabricated structural systems and components, and in production of typed structural designs and details. In this area integrated life cycle design can be used on a large scale. In the design of individual buildings, integrated life cycle design probably will only be partially used, but this part of the procedure can be applied through model designs for clients, architects and structural designers. In this way the application in ordinary cases can be simplified in order to avoid too much effort and cost, even though the design cost is marginal in comparison to the entire life cycle cost of the structures.

6.2 Further research needs

New basic knowledge will be needed concerning materials and structures, especially as regards environmental burdens, hygrothermal behaviour, durability and service life of materials and structures in varying environments. Data for lifetime costs especially with regard to maintenance and repair are needed. Structural design methods will have to be further developed that are capable of life cycle design, multiple analysis decision-making and optimisation. Recycling design and technology demand further research in design systematics, recycling materials and structural engineering. The knowledge obtained will have to be put into practice through standards and practical guides.

The creation of new types of materials and structures, in which the properties can be tailored separately for each specific need, is of vital importance. Both strong and soft solutions must be sought, depending on the specific life cycle requirements. New creative innovations for applications of by-products and recycling materials from industry and general consumption are also needed.

Appendix 1

List of international and national regulations, codes and standards

ISO standards

ISO 14001:1996, Environmental management systems. Specification with guidance for use.

ISO 14004:1996, Environmental management systems. General guidelines on principles, systems and supporting techniques.

ISO 14010:1996(E), Guidelines for environmental auditing – General principles.

ISO 14011: 1996, Guidelines for environmental auditing – Audit procedures – Auditing of environmental management systems.

ISO 14012: 1996, Guidelines for environmental auditing Qualification criteria for environmental auditors.

ISO DIS 14040, Environmental management – Life cycle assessment – Inventory analysis.

ISO 14050, Environmental management – Terms and definitions.

ISO/CD 15686–1, Guide for service life design of buildings. Part 1 – General principles.

ISO 6240–1980, Peformance standards in building – Contents and presentation.

ISO 6241–1984, Performance standards in building – Principles for their preparation and factors to be considered.

ISO 6707–1:1989, Glosary of terms.

ISO 7162–1992, Performance standards in building – Contents and format of standards for evaluation of performance.

ISO 9699–1994, Performance standards in building – Checklist for briefing – Contents of brief for building design.

National regulations and standards

Finland

Finnish Association of Civil Engineers (2001), *Lifetime structural engineering* (in Finnish). RIL 216–2001, Finnish Association of Civil Engineers, Helsinki 2001.

Germany

Regulations

Abfallbestimmungs-Verordnung (3 April 1990).
Abfall- und Reststoffüberwachungs-Verordnung (3 April 1990).

Bauabfall-Verordnung (10 January 1994).
Bundes-Immissionsschutzgesetz (14 May 1990).
Ökologische Durchführungsverordnung zur Bauordnung für das Saarland (ÖVO-LBO).
Reststoffbestimmungs-Verordnung (3April 1990).
Verpackungs-Verordnung (12 June 1991).
Zielfestlegung der Bundesregierung (5 November 1992).

Standards

(ATV) -DIN 18299. Die Verdingungsverordnung für Bauleistungen VOB. (Dez.1992) Teil
C. Allgemeine Technische Vertragsbedingungen für Bauleistungen.
DIN ISO 14001. Umweltmanagement.
DIN ISO 14010. Umweltaudit.
DIN ISO 14011–1. Umweltaudit–Auditverfahren.
DIN ISO 14012. Qualitetskriterien für Umweltauditoren.
DIN 33926: 1996–02. 'Umweltmanagement Produktbezogene Ökobilanzen'. *Standard-
berichtsbogen* DIN 33926, (01.02.96–31.10.96).

Guidelines

VDI 1000. Richtlinienarbeit. Grundsätze und Anleitungen. (Establishing of Guidelines.
Principles and Procedures). VDI-Richtlinien. Verein Deutscher Ingenieure. Oktober
1981.
VDI 2220 Produktplanung.
VDI 2221 Methodik zum Entwickeln und Konstruieren technischer Systeme und Produkte.
VDI-Richtlinien. VDI-Gesellschaft Entwicklung Konstruktion Vertrieb Ausschub
Methodisches Konstruieren. VDI-Handbuch Konstruktion. Verein Deutscher
Ingenieure. Düsseldorf Mai 1993
VDI 2222 Konstruktionsmethodik Methodisches Entwickeln von Lösungsprinzipien. VDI-
Gesellschaft Entwicklung Konstruktion Vertrieb. Ausschub Konstruktionmethodik.
Untcraussschub Mcthodischcs Entwickeln von Lösungsprinzipien. VDI-Handbuch
Konstruktion. VDI-Handbuch Getriebetechnik I. VDI/VDE-Handbuch Mikro- und Fein-
werktechnik. Juni 1997.
VDI 2225 Konstruktionsmethodik. Technisch-wirtschaftliches Konstruieren. Anleitung
und Beispiele. VDI-Gesellschaft Konstruktion und Entwicklung. Fachbereich Konstruk-
tion. VDI-Handbuch Konstruktion. VDI/VDE- Handbuch Feinwerktechnik. VDI-
Handbuch Betriebstechnik. Teil 1. Verein Deutscher Ingenieure. April 1977.
VDI 2232 Mcthodische Auswahl fester Verbindungen; Systematik, Konstruktionskataloge,
Arbeitshilfen.
VDI 2243. Konstruieren recyclinggerechter technischer Produkte. Grundlagen und
Gestaltungsregeln. VDI-Gesellschaft Entwicklung Konstruktion Vertrieb. Ausschub
Recyclinggerechte Produktgestaltung. VDI- Handbuch Konstruktion. Verein Deutscher
Ingenieure Oktober 1993.
VDI (1977) 'Material- und energiesparende sowie recyclinggerechte Gestaltung durch
methodisches Konstruieren', *VDI-Berichte* 277.
VDI (1991) 'Recycling – eine Herausforderung für den Konstrukteur', *VDI-Berichte* 906
VDI (1977) 'Optimale Rohstoffnutzung – eine Aufgabe für den Ingenieur', *VDI-Berichte*
277.
Weege, R.-D. (1981) *Recyclinggerechtes Konstruieren*. Düsseldorf: VDI-Verlag.

Woite, B. (1992) 'Recyclinggerechtes Konstruieren von Kunststoffbauteilen – Ansätze zur Normung', *DIN-Mitteilung* 71(2) 90–99.

Netherlands

Basis pakket duurzaam bouwen, Ministerie van VROM, Den Haag, 1995.

Beleidsverklaring Milieutaakstellingen Bouw 1995, Milieuberaad Bouw, June 1993.

NEN 7340:2000 NL. Uitloogkarakteristieken van vaste grond- en steenachtige bouwmaterialen en afvalstoffen – Karakteriseringsproeven – Algemene aanwijzingen (Leaching characteristics of solid earthy and stony building and waste materials – Characterization tests – General instructions).

NEN 7300:1997 NL. Uitloogkarakteristieken van vaste grond- en steenachtige bouwmaterialen en afvalstoffen – Monsterneming – Algemene aanwijzingen (Leaching characteristics of solid earthy and stony building and waste materials – Sampling – General instructions).

NEN 7310:1995 NL. Uitloogkarakteristieken van vaste grond- en steenachtige bouwmaterialen en afvalstoffen – Monstervoorbehandeling – Algemene aanwijzingen (Leaching characteristics of solid earthy and stony building and waste materials – Sample pretreatment – General instructions).

NEN 7320:1997 NL. Uitloogkarakteristieken van vaste grond- en steenachtige bouwmaterialen en afvalstoffen – Bepaling van het gehalte van anorganische componenten – Algemene aanwijzingen (Leaching characteristics of solid earthy and stony building and waste materials – Analysis of inorganic components – General instructions).

NEN 7330:2001 NL. Uitloogkarakteristieken van vaste grond- en steenachtige bouwmaterialen en afvalstoffen – Bepaling van het gehalte van organische componenten – Algemene aanwijzingen (Leaching characteristics of solid earthy and stony building and waste materials – Analysis of organic components – General instructions).

Regulations

Declaration on prohibition of dump of building and demolition waste.

Decision on which waste materials are considered 'dangerous' and regulation on the manufacturing of this dangerous waste.

Decision on prohibition of manufacturing of asbestos.

Decision on prohibition of CFC production and application.

Decision on prohibition of Cadmium production and application.

Guidelines

Declaration on avoiding use of tropical wood.

Agreement on avoiding use of tropical wood.

Declaration on enlarging the recycling of building and demolition waste.

Declaration on reduction of the use of hydrocarbon.

Policy declaration on reducing the application of PACs.

Policy declaration on reducing the risk of radiation.

Energy conservation *regulations*

Buildings Decree: the Energy Performance Standard.

Quality improvement *regulations*
Buildings Decree: Health in Indoor Environment.

* noise reduction
* moisture reduction
* discharge of rain and waste water
* ventilation
* daylight admission

Switzerland

Regulations

GSchG Gewässerschutzgesetz (1991).
LRV Luftreinhalte-Verordnung (1985).
StoV Verordnung über umweltgewährdende Stoffe (1986).
TVA Technische Verordnung über Abfälle (1990).
USG Umweltschutzgesetz (1983).
VVS Verordnung über den Verkehr mit Sonderabfällen (1986).

Standards and guidelines

NPK 117 D/95. Normpositionen-Katalog der Schweizer Bauwirtschaft Abbruch und
   Demontage (Schweizerische Zentralstelle für Baurationalisierung) .
NPK 211 Erdarbeiten (Schweizerische Zentralstelle für Baurationalisierung) .
NPK 212 Baugrubenaushub (Schweizerische Zentralstelle für Baurationalisierung).
NPK 221 D/96. Normpositionen-Katalog der Schweizer Bauwirtschaft
   Fundationsschichten und Materialgewinnung. Schweizerische Zentralstelle für
   Baurationalisierung.
NPK 311 Baumeisteraushub (Schweizerische Zentralstelle für Baurationalisierung).
1504 Richtlinien zur Verhütung von Unfällen bei der Ausführung von Abbrucharbeiten
   Grundlagen und Checklisten für die Erstellung von 'Integralen Sicherheitsplänen im
   Bauwesen', SUVA-Bau (in Vorbereitung).

Norms

Schweizer Norm SN 640 740. Recycling von Bauschutt. Verwertung von Bauschutt,
   Allgemeines. Vereinigung Schweizerischer Strassenfachleute. Mai 1993. 11 p.
Schweizer Norm SN 640 741. Recycling von Bauschutt. Verwertung von Ausbauasphalt.
   Vereinigung Schweizerischer Strassenfachleute. Mai 1993. 14 p.
Schweizer Norm SN 640 742. Recycling von Bauschutt. Verwertung von Strassenaufbruch.
   Vereinigung Schweizerischer Strassenfachleute. Mai 1993. 12 p.
Schweizer Norm SN 640 743. Recycling von Bauschutt. Verwertung von Betonabbruch.
   Vereinigung Schweizerischer Strassenfachleute. November 1993. 10 p.
Schweizer Norm SN 640 744. Recycling von Bauschutt. Verwertung von Mischabbruch.
   Vereinigung Schweizerischer Strassenfachleute. April 1994. 8 p.
SIA-Dokumentation D 093 Deklarationsraster für ökologische Merkmale von Baustoffen.
   1992.

SIA-Dokumentation D 0122 Ökologische Aspekte des Bauens. Mai 1995.
SIA-Dokumentation D 0123 Hochbaukonstruktionen nach ökologischen Gesichtspunkten. Schweizerischer Ingenieur- und Architekten-Verein. September 1995.
SIA 162/4 Recyclingbeton (in Vorbereitung). Schweizerischer Ingenieur- und Architekten-Verein.
SIA 203 Deponiebau (in Vorbereitung). Schweizerischer Ingenieur- und Architekten-Verein.
SIA 430 Entsorgung von Bauabfällen bei Neubau-, Umbau- und Abbrucharbeiten. SIA Empfehlung Ausgabe Schweizer Norm Bauwesen 509 430. Schweizerischer Ingenieur-und Architekten-Verein. 1993. 16 p. ON

United Kingdom

Standards

BS 6543: 1985. Guide to the use of industrial by-products and waste materials in building materials in building and civil engineering.
BS 7543: 1992. Guide to durability of buildings and building elements, products and components. British Standards Institution, London.
BS 7750: 1992. Specification for Environmental Management Systems
Halliday, S.P. (1994) *Environmental Code of Practice for Buildings and their Services.* BSRIA: Bracknell.

United States of America

Standards

E-50.06.01. Life Cycle Assessment.
E-50.06.03. Standard Guide for Design, Specification, Construction and Operation of Residential Green Buildings.
E.50.06.05. Sustainable Harvested Wood.
E.50.06.06. Environmentally Preferable Cleaners/Degreasers.
E-50.06.08. Sustainable Site Planning.
E-50.06.09. International Standard on Energy Efficiency.
E-50.06.10. Source Separation of Recyclables in Commercial and Multi-Tenant Buildings
ASTM Standard guide for Conducting a Life Cycle Impact Assessment of Building Materials. Draft 1996.
ASTM Standard Guide for Conducting a Life Cycle Inventory of Building Products and Building Materials. Draft 1996.
ASTM Standard Guide for Design, Specification, Construction and Operation of Residential Green Buildings. Draft 1996.
ASTM Standard Guide for Environmental Life Cycle Assessment of Building Materials/ Products. Draft 1996.
ASTM Standard Guide for Evaluation of the Design and Construction of Green Commercial Buildings. Draft 1996.
ASTM Standard. International Standard for Calculating Pollution Taxes Eighth Draft. 5 October 1995.

ASTM E632 Standard Practice for Developing Accelerated Tests to Aid Prediction of the Service Life of Building Components and Materials, Annual book of Standards, Vol. 04.07, American Society for Testing and Materials, Philadelphia, PA.

ASTM E-632 Developing Accelerated Tests to Aid Prediction of Service Life of Building Components and Materials.

ASTM E 917–93. Standard Practice for Measuring Life-Cycle Costs of Buildings and Building Systems. ASTM E06.81 Building Economics.

ASTM E 917–94. Standard Practice for Measuring Life-Cycle Costs of Buildings and Building Systems.

ASTM E 1557–93. Standard Classification for Building Elements and Related Sitework- UNIFORMAT II.

Standard Guide for Conducting a Life-Cycle Inventory of Building Products and Building Materials.

Standard Guide for Environmental Life-Cycle Assessment of Building Materials and Products.

Standard Guide for Conducting a Life-Cycle Impact Assessment of Building Materials.

Standard Guide for Evaluation of the Design and Construction of Green Commercial Buildings.

Terminology for Life-Cycle Assessment of Building Materials.

Appendix 2
Guidebooks, certification and manuals

Ammar, C. and Longuet, M. (1978), *Belgian Requirements For Building Service Life, Durability of Building Materials and Components.* ASTM Special Technical Publication 691. American Society for Testing and Materials: Philadelphia, PA.

Eco-Quantum (1996) *Design of a Calculation Method for the Quantitative Determination of the Environmental Impact of a Building: Final Report* (in Dutch) W/E Consultants Sustainable Building: Gouda and IVAM Environmental Research: Amsterdam.

Kortman, J.G.M. and van Roekel, A.L.W. (1996), *Environmental Regulations for Building Materials in Eight Surrounding States: A Study Commissioned by the Dutch Ministry of Housing, Spatial Planning and Environment (VROM).* IVAM Environmental Research, University of Amsterdam: Amsterdam.

Kortman, J.G.M. and van Ewijk, H. (1998) 'Presentation of Eco-Quantum, the LCA-based computer tool for the quantitative determination of the environmental impact of buildings'. *Materials and Technologies of Sustainable Construction: Construction and the Environment* CIB World Building Congress Gävle, Sweden 7–12 June.

Mak, J., Anink, D., Knapen, M., Kortman, J.G.M. and van Ewijk, H. (1997) 'Eco-quantum development of LCA-based tools for buildings', *Proceedings Second International Conference Buildings and the Environment*, Paris, 9–12 June.

RILEM Recommendations TC 71–PSL (in conjunction with CIB W80). *Systematic methodology for service life prediction of building materials and components.*

RILEM Recommendation. 'Systematic methodology for service Life prediction of building materials and components', *Materials and Structures*, 22, 385–392.

Denmark

Boligministeriet Byøkologi bygninger og bolige, Handlingsplan maj 1995.

Data fra offentlige registre og geografiske informationssystmer, GIS.

Håndbog I miljørigtig projektering. Bind 1, Metodebeskrivelse, Vejledninger. Publikation 121, høringsudgave, april 1997. BPS-centret, DTI Byggeri.

Håndbog I miljørigtig projektering. Bind 2, Miljødata, Eksempler. Publikation 121, høringsudgave: april 1997. DTI Byggeri.

Heino, E., Erat, B. Bygg- och rivningsavfall in Norden. NKB Utskotts- och arbetsrapporter 1996:07. Nordiska kommittén för byggbestämmelser, NKB. Arbetsgruppen för ekologiskt byggande. Nordiska ministerrådet. 50 p.

Livscyclus-basert bygningsprojektering,Edb-verktøj til energi og emissionsberegning, SBI medelelse 110, 1995.

Miljøanalysemodel for byggeri. Nr. 11 1994. Arbejdsrapport fra MiljÆstyrelsen. MiljÆministeriet Miljøstyrelsen.
Miljøriktig projektering:
Miljøhåndbog.
Miljødata for byggmaterialer, Delprojekt 3.
Miljøchecklister.
Minimering af ressouceforbrug, energiforbrug og miljøforurening I byggeri. Rammeplan for forsøg inden for Byggeri och Økologi Bygge- og Boligstyrelsen 1989.
Planlaeging av bygge- og anlaegesaffald, DEMEX maj 1996, Miljøstyrelsen og Københavns kommune.

Germany

Arbeitsgemeinschaft für zeitgemässes Bauen. e.V., Kiel (1992), 'Ökologisches Bauen, nachdruck Heft I, II, III. Umweltverträgliche Baustoffe'. *Mitteilungsblatt* der, vol. 189, 69 p.

Altop (1995) *Das alternative Branchenbuch*. Altop Verlags- und Vertriebsgesellschaft für umweltfreundliche Produkte mbH, Gotzinger Str 48, 81371 München.

'Recyclinggerechte Bauweisen im Innenausbau', in Bredenbals, B., and Willkomm, W. *Abfallvermeidung in der Bauproduktion. Bauforschung für die Praxis*, vol. 6, 197 p.

Bredenbals B., Willkomm W. and Weber II. Recyclinggerechte Bauweisen im Innenausbau. Informationszentrum RAUM und BAU, Stuttgart 1993, [BRE93].

Bredenbals, Barbara, and Willkomm, Wolfgang, (1994), Recycling bei Sanierungs-massnahmen. Kurzbericht, Kurzberichte aus der Bauforschung (1994) Jg.35, Nr.9, S.533–540, Anmerkungen: Englischer Kurzbericht im IRB vorhanden. Ausfuehrlicher Bericht: 208 S., abgeschlossen 03/1994, Bezug bei IRB Verlag, Best.-Nr. F 2250, ISSN: 0343–1118.

Bredenbals, B., Willkomm, W. and Weber, H. (Projektleiter) (1994) Recycling bei Sanierungsmassnahmen. Abschlussbericht. Stuttgart (Deutschland, Bundesrepublik): IRB Verlag. Mrz 1994, 208 S.

Bund Deutscher Architekten (BDA) (1994), *Umwelt-Leitfaden für Architekten* (Hrsg.) Ernst & Sohn, Berlin, 209 p.

Bredenbals, B., and Willkomm, W. (1993) *Kontaminierte Bauteile in Hochbau – Vermeidung, Erkennung und Behandlung*, RKW Schrift 5.2.2.07. RKW, Eschborn, 127 s. ISBN 3–926984–09–0.

Bredenbals, B. and Willkomm, W. (1994) *Abfallvermeidung in der Bauproduktion. Bauforschung für die Praxis*, vol. 6. 197 s. IRB Verlag, Stuttgart, ISBN 3–8167–4205–X.

Bund Deutscher Architekten (BDA) (1994) *Umwelt-Leitfaden für Architekten* (Hrsg.) Ernst & Sohn, Berlin, 209 p.

Bundesministers für Raumordnung, Bauwesen und Städtebau (1993) Bau- und Wohnforschung, Recyclinggerechte Bauweisen im Innenausbau F 2212. *Bauforschungsberichte*. IRB Verlag. 144 p.

Greiff, W. (1991) 'Ökologischer Mietwohnungsbau', in Bredenbals, B. and Willkomm, W. *Abfallvermeidung in der Bauproduktion. Bauforschung für die Praxis*, vol. 6. 197 p.

Haefele, G., Oed, W. and Sambeth, B.M. (1996) *Baustoffe und Ökologie, Bewertungskriterien für Architekten und Bauherren*, Bund Deutscher Architekten, Landesverband Baden-Württenberg, 373 p.

Handwerkskammer Hamburg (1995) Arbeitsgemeinschaft umweltverträgliches Bauprodukt e.V., München. Umweltbundesamt: Ökobilanzen für Produkte. Texte 38/92. Berlin 1992.

Hösel, G. and Freiherr. v. Lersner, H. (1995) Recht der Abfallbeseitigung des Bundes und der Länder. Kommentar zum Abfallgesetz, Nebengesetze und sonstiger Vorschriften. Erich Schmidt Verlag, Berlin, 3 Bände, 1. Aufl. 1972, jährl. erg. Loseblattsammlung.

Kreislaufwirtschaft im Baugewerbe. 4. Weimarer Fachtagung über Abfall- und Sekundärrohstoffwirtschaft. Schriftenreihe 04 der Professuren Abfallwirtschaft und Siedlungswasserwirtschaft, Professur Aufbereitung von Baustoffen und Wiederverwertung. Bauhaus-Universität Weimar 1996.

Kreislaufgerechtes Bauen im Massivbau. Darmstädter Massivbau-Seminar. Band 18. Darmstadt 1997.

RAL-RG 501/1 'Recycling-Baustoffe für den Strassenbau'. Deutsches Institut für Gütesicherung und Kennzeichnung e.V., Bonn 1985.

Recyclinggerechte Bauweisen im Innenausbau F2212. Bau- und Wohnforschung. Bauforschungsberichte des Bundesministers für Raumordnung, Bauwesen und Städtebau.

Willkomm, W. (1996) Recyclinggerechtes Konstruieren im Hochbau. (RKW-Verlag) Eschborn.

Willkomm, W. (1993) Recyclinggerechtes Konstruieren im Hochbau: Recycling-Baustoffe einsetzen, Weiterverwertung einplanen. Köln: Verlag TÜV Rheinland.

Bredenbals, B. (1993) Verband Deutscher Baustoff-Recycling-Unternehmen e.V. (Hrsg.) Recycling, Rückbau und Umweltgerechte Baustellenentsorgung. Bonn, 1993.

ks. myös: http://www.bauwesen.fh-kiel.de/bauwesen/fbb/infosysbaunu/Literatur/lit-21.html

Eibl, J. and Walther, H.-J. (1996) 'Umweltgerechter Rückbau und Wiederverwertung mineralischer Baustoffe'. *Deutscher Ausschuss für Stahlbeton*, vol. 462. Berlin. 181 p.+ 11 p.

Statistisches Bundesamt (1987) 'Abfallbeseitigung im produzierenden Gewerbe', Fachserie 19, Reihe 1.2, Stuttgart, in Bredenbals, B. and Willkomm, W. *Abfallvermeidung in der Bauproduktion. Bauforschung für die Praxis*, vol. 6, 197 p.

Umweltbundesamt (Hrsg.) (1993) *Leitfaden zum ökologisch orientierten Bauen*, 2. Auflage. Verlag C.F. Müller, Karlsruhe, 86 p.

Japan

Architectural Institute of Japan (1993) *Principal Guide for Service Life Planning of Buildings*. English edition. Architectural Institute of Japan: Tokyo.

Netherlands

Boonstra, C., Anink, D. and Mak, J. (1996) *Handbook of Sustainable Building. An Environmental Preference Method for Selection of Materials for Use in Construction and Refurbishment*, James & James: London.

Norway

Berge, B. De siste sykehus 1990. Universitetsforlaget, Oslo.

Berge, B. Bygningsmaterialenes Økologi. Universitetsforlaget, Oslo 1992 .

Berge, B. Bygningsmaterialer for en en baerekraftig utvikling. NKB Utskotts- och arbetsrapporter 1995:07, nordisk komité for bygningsbestemmelser, NKB, Arbeidsgruppen for oekologisk bygging.

Brukerhåndbok for boliger. Verktøykasse for en sunnere by, Redskap 6, Bergen kommune. 1995. 66 p.

En sunnere by? Utgift av Sunnere By, Bergen kommune, September 1995.

Flytting av bygninger (Byggforskserien: Byggforvaltning 700.126; 1995).

Håndbok for planlegging av boliger og tilltak for eldre I etablerte bomiljø. Verktøykasse for en sunnere by, Redskap 7, Bergen kommune.1995. 42 p.

Kommunal styring av bygg- og anleggsavfall – Statusrapport september 1995 (Oslo Renholsverk 1995).

Kommunal styring av bygg- og anleggsavfall – delrapport 1995 (Oslo renholdsverk/Plan- og bygnings-etagen 1995).

Kvalitetssikring av en miljøvennlig byggeprocess. 8 p.1995. Verktøykasse for en sunnere by, Redskap 4, Bergen kommune.

Livsløpsvurdering av bygningsmaterialer (Byggforskserien: Byggdetaljer 470.101, sending 1 1995).

Miljøsanering ved rivning og rehabilitering (Byggforskserien: Byggforvaltning 700.802; Sending 1–1995)

Miller, F., Reite, A. Levende hus- om miljø- og ressursvenlig bygging. TI-forlaget, Oslo. 1993.

Planleggning og bygging med lite avfall (Byggforskserien: Byggdetaljer 501.101, sending 2–1994).

Reduksjon og håndtering av byggavfall (Byggforskserien: Byggdetaljer 501.105, Sending 1–1994).

Ren och ryddig byggeprosess (Byggforskserien: Byggdetaljer 501.107; Sending 1–1995).

Rongen, A. Byøkologi.

Sjekkliste for sunnere planleggning (1995) Verkt/Eykasse for en sunnere by, Redskap 1, Bergen kommune. 20 p.

Utbyggeravteler for miljÆforbedret boligbygging. 20 p.

Verktøykasse for en sunnere by, Redskap 3, Bergen kommune. 1995.

Varefakta for boliger 20 p. Verktøykasse for en sunnere by, Redskap 5, Bergen kommune. 1995. 20 p.

Sweden

(1995), Producentansvar för byggvaror- kretsloppsanapassad rivning, Karlskrona, Boverket.

Bergh, Å. and Lundquist, B. Planera för återvinning. Solna. Svensk byggtjänst och REFORSK FoU-rapport 125.1995.

Blix, K. (1995), Producentansvar för byggvaror i Sverige. Ekoark, nr 1.

Gustafsson, M. and Sahlin, F., Datorstödd riviningsplan. Förstudie. Byggmästareföreningen Väst. FoU-Väst, Rapport 9504, Göteborg 1995.

Gustavsson and Wibom. Bygga miljövänligt-hur då? Tekniska högskolan, ingenjörsskolan, Haninge 1994.

Johansson, B. (1995) Bygg- och rivningsmaterial I kretsloppet.. Dagsläge och kunskaps-behov.Skrift T7:, Byggforskningsrådet.

Sigfrid, L. Avfall på bygget – en praktisk hjälpreda. Svensk Byggtjänst, Solna 1995.

Sternberg, H. Ekobygg- Produktguide för sunda och miljöanpassande hus 1995–96. Ekokultur förlag, Falun 1995.

Sustainable Building, An Ecocycle System in building and Construction. A Proposal to the Foundation for Strategic Environmental Research MISTRA. Chalmers University of Technology, Lund Institute of Technology, Royal Institute of Technology, Swedish Testing and Research Institute

Thormark, C. (1995) Byggande för minskat avfall och ökad återanvänding – en kunskapsöversikt. Lunds tekniska högskola, Institutionen för byggnadsfunktionslära.

Thormark, C. Återbygg – Möjligheter och problem med återvinning av byggnadsmaterial. Lunds tekniska högskola, Institutuionen för byggnadsfunktionslära, rapport TABK 95/3028 1995.

Switzerland

Richter, K. and Sell, J. (1992) 'Ökobilanzen von Baustoffen und Bauprodukten aus Holz, Zusammenfassung erster Erkenntnisse'. *Forschungs + Arbeitsberichte*, 15/25, Eidgenössiche Materialprüfungs- und Forschungsanstalt EMPA: Dübendorf.

United Kingdom

BRE (1981) *Waste of Building Materials*, Digest 247, BRE: Watford.

BRE (1982) *Materials Control to Avoid Waste*, Digest 259, BRE: Watford.

Snook, K., Turner, A. and Ridout, R. (1995) 'Recycling waste from the construction site'. *The Chartered Institute of Building Site Environmental Handbook*. The Chartered Institute of Building: Ascot.

CIRIA (1994) *Environmental Handbook for Building and Civil Engineering Projects: Volume 1 – Design and Specification*. Special publication 9. CIRIA: London.

CIRA (1994) *Environmental Handbook for Building and Civil Engoineering Projects: Volume 2 – Construction Phase*. Special publication 98. CIRIA: London.

Thomas, R. (1996) *Environmental Design*. The Timber Research and Development Association (TRADA) and E & FN Spon: London.

USA

American Forest & Paper Association (1996) *National Wood Recycling Directory*. American Forest & Paper Association: Washington, DC.

Demkin, J.A. (ed.) (1996) *Environmental Resource Guide*. John Wiley & Sons Inc: New York.

Boston Society of Architects (1992) *The Sourcebook for Sustainable Design: A Guide to Environmentally Responsible Materials and Processes*. Architects for Social Responsibility: Boston, MA.

Boulanger, J. (ed.) (1995) *The Official Recycled Products Guide*. Recycling Data Management Corporation: Ogdenburg.

Center for Resourceful Building Technology (1995) *Guide to Resource Efficient Building Elements*, 5th edn. Center for Resourceful Building Technology: Missoula.

City of Austin (1996) *Green Builder Program 499–star Directory*. City of Austin: Austin.

City of Los Angeles Integrated Solid Waste Management Office (1993) *Wood You Recycle? A Guide to Wood Waste Re-Use and Recycling in the L.A. Area*. Recycling Resources Series. City of Los Angeles Integrated Solid Waste Management Office: Los Angeles.

Environmentally Preferable Products Proposed Guidance. EPA/744–F-94–002. 1995. US Environmental Protection Agency, Pollution Prevention and Toxics: Washington, DC.

EPA-452/R-95–002. (1995) 'Life-Cycle Impact Assessment: A Conceptual Framework, Key Issues, and Summary of Existing Methods'.

Innovative Waste Management (1993) *Construction Materials Recycling Guidebook: A Guide to Reducing and Recycling Construction and Remodeling waste*. Innovative Waste Management: Minnesota.

L.A. Network (1995a) *A Resource Guide to Recycled-content Construction Products*. Buy Recycled series. L.A. Network: Los Angeles.

L.A. Network (1995b) *Construction and Demolition Waste Recycling Guide*. Recycling Construction and Demolition Waste in the Los Angeles Area, Issue 13, L.A. Network: Los Angeles.

LEED Buildings (1996) Environmental Building Rating System Criteria. Fourth Ballot Draft. U.S. Green Building Council. [unable to put this reference into style, please advise].

Minnesota Office of Environmental Assistance (1993) *A Business Guide to Recycling Market Development*. Minnesota Office of Environmental Assistance: Minnesota.

Minnesota Office of Environmental Assistance (1995a), *Minnesota Recycled Product Directory: A Guide to Products Containing Recycled Materials Made by Minnesota Companies*. Minnesota Office of Environmental Assistance: Minnesota.

Minnesota Office of Environmental Assistance (1995b), *Problem Materials Plan II*. Minnesota Office of Environmental Assistance: Minnesota.

Norris, G.A. and Marshall, H.E. (1995) *Multiattribute Decision Analysis Method for Evaluating Buildings and Building Systems*. NISTIR 5663, U.S. Department of Commerce, Technology Administration. Building and Fire Research Laboratory. National Institute of Standards and Technology: Washington, DC.

Public Technology Inc. and U.S. Green Building Council (1993) *Local Government Sustainable Buildings Guidebook: Environmentally Responsible Building and Management*. Public Technology Inc. and U.S. Green Building Council: Washington, DC.

Public Technology Inc. and U.S. Green Building Council (1996) *Sustainable Building Technical Manual: Green Building Design, Construction and Operations*. Public Technology Inc., and U.S. Green Building Council: Washington, DC.

Stafford Harris Inc. (1996), *The Harris Directory of Recycled Content Building Materials*. (database) Stafford Harris Inc.: Port Townsend.

Glossary

The terms in the glossary are listed alphabetically under the following headings:

- Life cycle and lifetime
- Serviceability and service life
- Reliability and performance
- Durability
- Management and maintenance
- Systems
- Personnel
- Methods

Life cycle and life time

Integrated life cycle (lifetime) design   Selecting and producing descriptions for structures and their materials to be used in design, which fulfil the specified requirements of financial costs, human requirements (safety, health, comfort), ecology (environmental costs), culture and social needs, throughout the complete life cycle of the structure. Integrated structural design is the synthesis of mechanical design, durability design, physical design and environmental design.

Life cycle   The consecutive and interlinked stages of a facility or structure, from the extraction or exploitation of natural resources to the final disposal of all materials as irretrievable wastes or dissipated energy.

Life cycle cost   Total cost of a structure throughout its life, including the costs of planning, design, acquisition, operations, maintenance and disposal, minus any residual value.

Lifetime   The time period from start of use of a facility or structure until a defined point in time.

> **Design lifetime**   A specified period of the life time, which is used in design calculations.

Serviceability and service life

Serviceability   Capacity of a structure to perform the service functions for which it is designed.

Service life   Period of time after installation during which a facility or its parts meet or exceed the performance requirements.

Characteristic service life A time period which the service life is expected to exceed with a specified probability, usually with 95% probability.

Design service life Used in design to provide a required probabilistic safety against falling below the target service life. Design service life is calculated by dividing the characteristic service life with lifetime safety factor. Design service life has to exceed the target service life.

Reference service life A service life forecast for a structure under strictly specified environmental loads and conditions for use as a basis for estimating service life.

Residual service life Time between present moment in time and the forecasted end of service life.

Target service life Required service life which a structure is expected or predicted to have, under a prescribed set of in-use conditions, with anticipated maintenance, but without any major repairs being necessary.

Service life design Preparation of the brief and design for a structure and its parts so as to achieve the desired design life, in order to control the usability of the structure and facilitate maintenance and refurbishment.

Reliability and performance

Factor method Modification of reference service life by factors to take account of the specific environmental loads and in-use conditions.

Failure Loss of the ability of a structure or its parts to perform a specified function.

    Durability failure Exceeding the maximum degradation or falling below the minimum performance parameter.

Failure probability The statistical probability of failure occurring.

Lifetime quality The capability of the facility to fulfil all requirements of the owner, user and society over the specified life time period.

Lifetime safety factor Coefficient by which the characteristic life is divided to obtain the design life.

Lifetime safety factor method or durability limit state method A statistically based deterministic method for calculating the required reliability level of the service life of structures for defined serviceability and ultimate limit states.

Limit state A state beyond a specified measure or performance.

    Durability limit state Minimum acceptable state of performance or maximum acceptable state of degradation, set with regard to a serviceability or ultimate limit state.

    Serviceability limit state State beyond specified service requirement(s) for a structure are no longer met.

    Ultimate limit state State beyond which collapse, or other similar forms of failure may occur.

Obsolescence Loss of ability of an item to perform satisfactorily due to changes in economic performance, human (safety, health, comfort), cultural or ecological requirements.

Performance Measure of how a structure responds to a certain required function.

Performance requirement or performance criterion Qualitative and quantitative levels of performance required for a critical property of a structure.

Risk Exposure to the possibility of loss, injury, or other adverse circumstance, e.g. structural failure or damage.

Durability

Condition   Level of critical properties of a structure or its parts, determining its ability to perform.

Condition model   Mathematical model for placing an object, module, component or subcomponent in a specific condition class.

Degradation   Gradual decrease in performance of a material or structure.

Degradation load   Any of the groups of environmental loads or mechanical loads.

Degradation mechanism   The sequence of chemical, physical or mechanical changes that lead to detrimental changes in one or more properties of building materials or structures when exposed to degradation loads.

Degradation model   Mathematical model showing degradation over time.

Deterioration   The process of becoming impaired in quality or value.

Durability   The capability of a structure to maintain minimum performance under the influence of environmental degrading loads.

Durability model   Mathematical model for calculating degradation, performance or service life of a structure.

Environmental load   Impact of environment onto structure, including weathering (temperature, temperature changes, moisture, moisture changes, solar effects etc.), chemical and biological factors.

Performance model   Mathematical model for showing performance over time.

Management and maintenance

Condition assessment   Methodology for quantitative measurement and visual inspection of the properties of a structure and its parts, and conclusions drawn from the results regarding the condition of the object.

Integrated lifetime management   Planning and control procedures implemented in order to optimise the economic, human, ecological and cultural conditions over the life cycle of a facility.

Maintenance   Combination of all technical and associated administrative actions during a structure's service life to retain it in a state in which it can perform its required functions.

M&R   Maintenance, plus any or all of the following: repair, restoration, refurbishment and renewal.

Recovery   A generic term for reuse, recycling or combustion of a structural system or its part.

Recycling   Use of materials of demolished structure or other product after treatment, as raw materials for new structures or other products.

Refurbishment or rehabilitation   Modification and improvements to an existing structure to bring it up to an acceptable condition.

Repair   Return of a structure to an acceptable condition by the renewal, replacement or mending of worn, damaged or degraded parts.

Restoration   Actions to bring a structure to its original appearance or state.

Reuse   Use of a structure or structural part after dismantling for the same purpose as its initial use.

Systems

Hierarchical system   A system consisting of some value scale, value system or hierarchy.
Facility or **object**   (e.g. bridge, tunnel, power plant, building) a basic unit of the network serving a specific function.
Material   Substance that can be used to form products.
Module or **assembly**   A part of an object, or a set of structural components which is designed and manufactured to serve a specific function or functions as a part of a system, and whose functional, performance and geometric relationships to the structural system are specified.
Modularised system   A system whose parts (modules) are autonomous in terms of performance and internal structure.
Network   Stock of objects (facilities, e.g. bridges, tunnels, power plants, buildings) under the management and maintenance of an owner.
Structural component   A basic unit of the structural system, which is designed and manufactured to serve a specific function or functions as a part of a module, and whose functional, performance and geometric relationships to the structural system are specified.
Structural system   A system of structural components which fulfil a specified function
Subcomponent   Manufactured product forming a part of a component.
System   An integrated entity which functions in a defined way and whose components have defined relationships and rules between them.

Personnel

Contractor   Person or organisation that undertakes, carries out or manages construction work. The contractor bids for a contract for a new building with information from manufacturers/suppliers. The contractor's representative on the building site is the site supervisor.
Designer   Person or organisation that prepares a design or arranges for any person under his control to prepare the design.
Manager   At take over the building is administered by a property manager who engages specialists to be responsible for proper maintenance inspections and/or to carry out the necessary maintenance.
Owner   Person or organisation for which structure is constructed.
Stakeholders   Owners, users, designers, contractors, industry sector/public interest organizations, and/or government agencies who have a connection with the structure in some role during the life cycle.
User   Person, organisation or animal which occupies a facility and has the responsibility for maintenance and upkeep of structural, mechanical and electrical systems of the building.

Methods

Allocation   The division of specified resources (financial and physical) into objects, projects and other actions on the network level.
Briefing   Statement of the requirements of a facility.
Service life planning   Preparation of the brief and design for a facility and its parts in order to optimise the required properties of the facility for owners and facilitate maintenance and refurbishment.

Index